The Architecture of the Wire

The Architecture of the Wire
Infrastructures of Telecommunication

Carlotta Darò

translated by Christian Hubert

THE MIT PRESS
Cambridge, Massachusetts London, England

Preface

Telephones (1995) by Christian Marclay is a video piece that consists of a montage of scenes from cult films, each showing a character communicating by telephone.

A sense of suspense builds up progressively, from the finger dialing the number, to the phone ringing, through various greetings such as "hello" or "yes," to a curious assemblage of snippets of conversations, silences, and the final "goodbye." The different emotional charges of the protagonists dealing with this sound machine give a powerful intensity to this portrait of the telephone in Hollywood cinema. While it is an omnipresent figure in our daily lives, in many films the telephone plays the role of the missing, absent, desired, or apprehended character, hidden behind a panoply of sound codes—the tones, the ringing, the voice transformed by electrical signals—and of objects embodying modern telecommunications—the phone booth, the receiver, the dial and keys of the telephone, right up to the cell phone.

This artist's montage is a kind of testimony to the vestiges of a more than one hundred-year-old system of machines and networks that was, at the time of its realization, heading for a major transformation due to the global diffusion of wireless telephony and portable devices. The sequence of images reveals the distinctive properties of its original function. The telephone offered the gift of ubiquity—a communication between two people that is possible in any given place at the same time. The telephone is always connected to a wire—wires that cross apartments, domestic spaces as well as public places. Its network is a watermark of modern architecture; the telephone separates oral communication from the rest of the body, generating its own codes of operation.

Almost fifty years ago, Ithiel de Sola Pool organized a series of seminars at the Massachusetts Institute of Technology to celebrate the centenary of the invention of the telephone.[1] The invention itself and its history were the object of study at the core of a series of more or less quantitative essays marking an important step in the field of the social sciences. The technical object was thus recognized as an element of social and cultural impact. Later studies adopted this approach attempting to broaden the analysis to the scale of networks and systems, beyond the invention of the telephone alone. They targeted a much more complex

Christian Marclay, *Telephones*, 1995, video, 7 min., 30 sec.
© Christian Marclay. Courtesy White Cube.

context, at once relational, societal, urban, architectural, political, economic, and technical.[2] Inspired by Emily Thompson's work on the soundscape of modernity, proposing a new cultural reading of the way in which acoustic science and technology transformed the architectural and urban production of the modern era—as well as certain social practices that stem from it, this study attempts an approach that is more strongly anchored in the materiality of the technical objects of telecommunications systems and their possible readings in the field of design.[3] It is grounded on the assumption that electrical technologies for sound transmission didn't develop autonomously from architectural and urban design. Conversely, through their physical constitution and functioning it becomes evident to which extent telecommunications networks were essentially a matter of design and how they were interwoven into modern planning. Electrical conducting systems were physically integrated into architecture and the city, and this condition had necessarily direct and mutual implications in terms of both the theoretical discourse and the project.

This book deals directly with the presence of cables, wires, and poles, as well as receptacles, switches, and conduits that are found inside built spaces. These objects allow us to interrogate architecture at several scales—from the landscape, to the domestic interior, to the individual space of a cabin—and through several themes. In a deliberate manner, the different parts of this work do not return to a "heroic" history of the inventions and to the exciting narratives surrounding their main actors (Bell, Edison, Berliner, Marconi), but they aim rather to reintegrate the narrative of an urban and rural transformation, the assimilation of or resistance to new objects, grids, and systems; the beliefs, the imaginaries, the symbols invoked by the public during the beginning of the so-called media revolution of modern times.

The actors of this history are manifold: engineers, company managers, but also workers, technicians, and above all users. It is an "anonymous" history that, by its plurality of axes, does not pay homage to a single person or invention.[4] Similarly, the material collected in this lineage of works refers mainly to the world of business and popular culture, all products of a creativity without an author, while seeking, from time to time, significant evidence of social and cultural change in action, both in art and in architecture.[5] In line with this tradition of anonymous histories, we speak today of a "material" history that is written from the "comfort devices" that came to structure the architectural project in the twentieth century.[6] It is to reach an ideal of comfort, inscribed in the architecture from its origin—with the image of fire as energy reserve—that modern architecture is filled with sheaths, wires, and switches, generating a complex technical and mechanical apparatus.

However, between the technical materiality of networks and the "culture of well-being" there are two opposing models of thought that come together—one based on quantitative measurement and the other on the non-objectifiable evolution of morals and thought—in a form of apparent oxymoron that is in fact the basis of any cultural reading of the history of techniques.[7]

In his *Esthétique industrielle* (Industrial aesthetics), Stanislaus von Moos mentions the wire and the power line among the objects that define the industrial culture of the Swiss territory, a "bric-a-brac of watches, tins, lace, mountain hotels, locomotives, typewriters and steam engines, carved Brienz knick-knacks and Bill model chairs."[8] He observes from the original setting of the mechanized nation of Switzerland, at the turn of the previous century, the existence of two intrinsic kinds of receptions related to these objects: on the one hand power lines as markers of an outraged landscape, and on the other hand as modern commodities, signs of progress, allowing economical and touristic developments. If these judgments are mainly based on an interpretation of telecommunications systems as agents of modernization, reflecting different responses to this, the author calls on aesthetic categories to address their physical presence within the territory.

Beyond the modern idealization of beauty based on the reduction of aesthetic categories to their simple value of use and efficiency coming from the specialization of technical knowledge, Georges Simondon's interpretation of beauty relates the technical object to a kind of cultural ecumenism, one that is largely relevant to this study.[9] Addressing pylons and cable, the author says: "A line of pylons supporting cables spanning a valley is beautiful, while the pylons, seen on the trucks that bring them in, or the cables, on the large rollers used to transport them, are neutral.... But it's not simply the technical object that's beautiful: it's the singular point in the world that the technical object embodies. It's not just the line of pylons that's beautiful, it's the coupling of the line, the rocks and the valley, it's the tension and bending of the cables: therein lies a silent, ever-continuing operation of technicality applied to the world."[10]

The sentiment of beauty is then born of a reticulation of thought that links several facets of the terrestrial realm. Technical objects are beautiful when they fit into a world, whether geographical or human. The aesthetic impression is then relative to the insertion, it is like "a gesture." Further on, Simondon says this about the telephone exchange and the switching lights on a telephone board:

> If the alignment of the panels in a telephone exchange is beautiful, it's not in itself, nor in its relation to the geographical world, for it could be anywhere; it's because these luminous lights, tracing multicolored, moving constellations from moment to moment, represent the real gestures of a multitude of human beings linked to one another by the interweaving of circuits. The telephone exchange is beautiful in action, because it is at every moment the expression and realization of an aspect of the life of a city and a region; a light is an expectation, an intention, a desire, an imminent piece of news, a ringing that we won't hear but that will draw us away to another house. This beauty is in the action, not only instantaneous, but also made up of the rhythm of peak hours and night hours. The telephone exchange is beautiful not because it is an object, but because it is a key point in collective and individual life.[11]

Thus, not only do infrastructures such as telecommunications systems connect individuals through power lines and technical objects, but also their proper functioning offers ways to observe the complexity of social exchanges, embracing a collective and public dimension, through different territorial and architectural realities, human habits, and rhythms. This book is about the multiple articulations of the wire.

1 *Matter: Anatomy of the Line*

Several other networks already existed when the telephone was first invented. Some of them, like the telegraph, were already using wires, cables, and standing poles. In July 1837, William Fothergill Cooke and Charles Wlieatstonc demonstrated the first telegraph link along about one-mile-long electric wire north of London. The following year, Cooke and Wheatstone opened the first commercial line from Paddington station to West Drayton, which was initially designed to be adjacent to the railway lines, in order to ensure its safety. The invention of the telephone, or oral communication at a distance, dates back to 1876. Inspired by the telegraphic network, the telephone infrastructure developed in parallel with the electrical transmission system, and rapidly grew during the two and three decades that followed. In the United States, the telephone network was rapidly commercialized by its inventor's company, the Bell Telephone Company, and it was there that the telephone network began an expansion and proliferation that would soon become global.

This book will follow the deployment and the effective presence of the line, starting from the raw texture of its materials and its forms of distribution, and consider the network in the landscape, as opposed to a recent tradition that assimilates modern networks to flows, and proceeds to forms of abstraction of the functioning of these systems. This approach celebrates, for example, the "death of distance" or the "end of geography," while adopting old biological metaphors—such as the bloodstream—to describe the movement of flows. More "materialist" approaches still restore a physical consistency to the field of geography.[1] Anna Tsing, echoed by Rania Ghosn, has asserted that globalization is concretized in a "sticky materiality" of practical encounters, which underscores the differences and inequalities of the interconnections that constitute society, otherwise known as the space of "frictions."[2] Starting out from an exploration of the physical friction of the objects, one can observe the complexity of social, political, economic, as well as aesthetic and cultural issues that underlie the physical presence of the network, in particular the telephone.

This sequence of chapters offers multiple perspectives on modern telecommunications objects, which are no longer seen as simple compositions of matter. These objects take on other meanings in relation to persons in their uses and history, as well as in the narratives or iconographic stories they generate. In Heideggerian terms, the infrastructures of telecommunications consist of assemblages of "things," as distinguished from "objects," because beyond their simple function, they absorb different types of "encounters" in the present time of their existence: from the human being to culture, philosophy, ideology, and politics.[3] They are like the pipelines defined by Darin Barney as "things" that are in the world, or else infrastructures that serve multiple media functions—"they contain, store, convey, conduct, transmit, connect, distribute, span." The lines of telecommunication also accomplish several forms of mediation. Like the pipelines whose existence is never neutral, and whose uses are "far from being simple lines on a map,"[4] the lines of telecommunication are not simple marks either. Thus, the lines conduct and transmit; they distribute and connect, but they also enable translation (of electricity into words). They speak and they sound. They themselves carry signifiers. And in the thick materiality of these networks, one finds traces of the cultures that built, adopted, exploited, and imagined them. Hence, by exploring the more circumscribed encounters around the wire and the cable, the post and the electric plug, the following chapters take on a form of coherence or global legibility.

Among several instructional books that appeared at the time of the first installation, William J. Hopkins's synthesis of the anatomy of the line, written for linemen and line planners, remains worthy of note.[5] The author was a professor of physics at the Drexel

Lineman with cable, n.d. Courtesy of AT&T Archives and History Center.

Institute of Art, Science, and Industry in Philadelphia. His manual was published in 1893 and subsequently went through several editions. The lineman—the man of the line, the repairman and installer of technical networks—was a new profession that would define a concrete and specialized relationship with the diffuse infrastructure of the line.

The author of the manual described in detail the principles of planning that aimed primarily at efficiency, but also at attractive appearance.[6] A line that works better, said the author, is also a line that looks better. Here two elements of the Vitruvian triad, *utilitas* and *venustas*, define the criteria for an aesthetic of infrastructure, which would soon interweave the field of architecture. This functionalist beauty is not simply an effect of good functioning, as in the case of modern functionalism. It contributes to the good functioning of the network itself since its topographical, harmonious insertion will guarantee its better preservation.

The functionality and solidity of the infrastructure also served as pledges for a network that from its outset was the source of danger and accidents. An 1877 litigation, following a violent storm that hit New York, ruled: "It [the telephone company] is bound to use reasonable care in the construction and maintenance of its line, so that no traveler shall be injured by it, and the amount of care must be proportioned to the amount of danger and the liability to accident." The line had to be able to weather a storm of "reasonable" intensity, but it was not expected withstand some extreme and unforeseeable weather event.[7]

This account draws on some of the main principles that define the form—the anatomy—of the telephone network, while acknowledging that despite the basic simplicity of its constituents, different technical and material developments have occurred in different countries over these almost one hundred fifty years.

Pole

The posts that support the overhead lines are preferably made of Canadian cedar, with a minimum diameter of seven inches in rural areas, and slightly more for urban lines. They should be stripped and shaved, and the joint for the cross arm should be cut before installation, which is done to a depth of about six feet into the ground. Only urban poles are treated with two coats of paint to preserve the wood. The size of the post varies according to the situation. In the city, it is preferred that posts be high enough to avoid obstructing the light or interfering with urban activity. The cross arm is made of Norwegian pine, covered with an insulating mineral paint. The arms are 4¼ by 3¼ inches thick and are generally ten feet long. Held by iron straps, they are fixed by pins. The poles are installed every one hundred or one hundred twenty feet with the arms coupled face-to-face, or back-to-back.

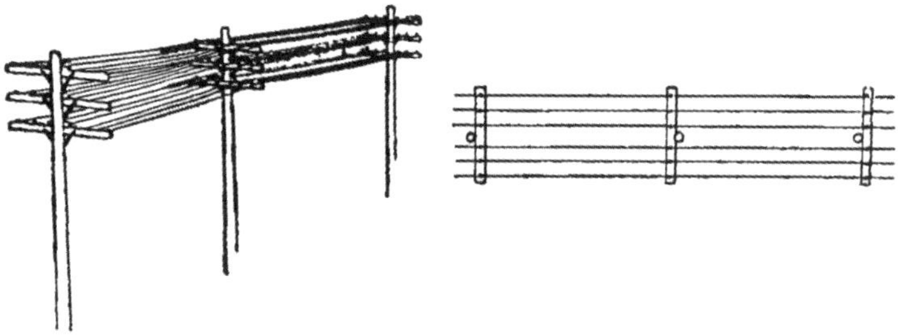

Positioning of the arms in a succession of posts. William J. Hopkins, *Telephone Lines and Their Properties* (New York: Longmans, Green, and Co., 1899), 6.

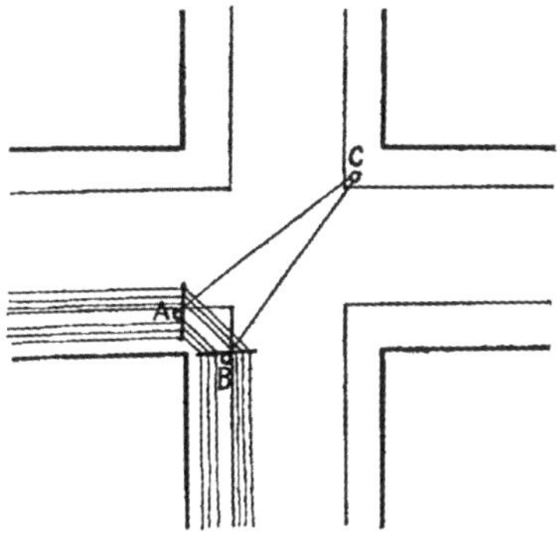

Methods of bracing of poles. William J. Hopkins, *Telephone Lines and Their Properties* (New York: Longmans, Green, and Co., 1899), 7.

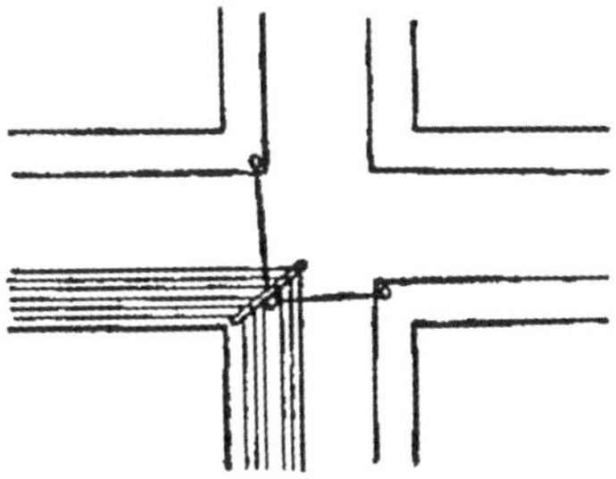

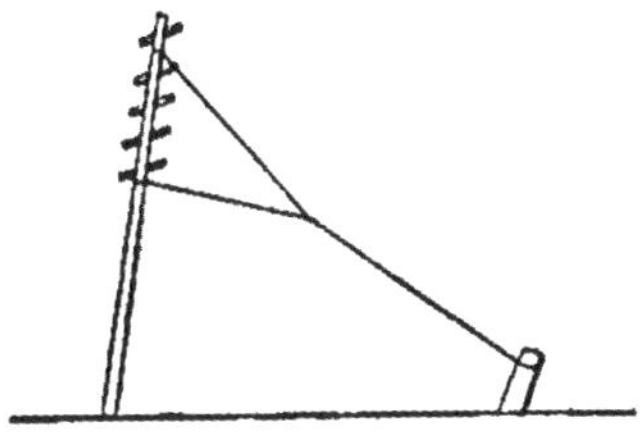

Methods of bracing of poles. William J. Hopkins, *Telephone Lines and Their Properties* (New York: Longmans, Green, and Co., 1899), 8.

This alternation avoids a chain of failures if a first pole falls due to an external accident. The tips of the poles in a line should be aligned as much as possible for maintenance and visual order. In the case of uneven topography, the posts must be cut to different heights to compensate for the unevenness of the ground. For posts that are at an angle, a line should be drawn in the opposite direction when possible, or otherwise they should be strongly braced to avoid their possible tilt. The post diameter should be larger, and the braces should pull the arm toward the post, not the other way around.

Wire

Galvanized iron was generally used for urban lines at first, as it had the advantage of strength and economy. But this material proved to be inadequate for electrical transmission, as it corrodes easily

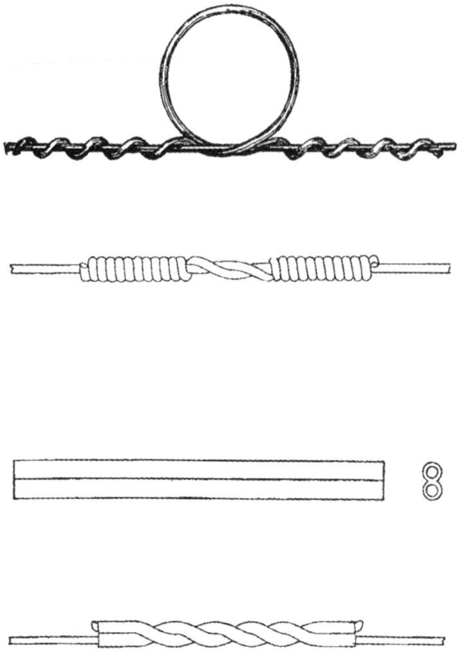

Different methods of connecting wires. William J. Hopkins, *Telephone Lines and Their Properties* (New York: Longmans, Green, and Co., 1899), 11.

due to the acids present in city air. Subsequently, pressed copper came to replace iron, as it was sufficiently strong and resistant to corrosion. The wire must be about 0.08 inch in diameter, in order to withstand possible weights and forces. A gray-green chloride patina forms over time and protects the inside of the wire from further outside influences. The tension of the wire varies according to the climate. It is tauter in cold climates, in order to support the weight of snowstorms and sleet. Since heat relaxes the copper, it is possible to span about one hundred twenty-five feet from one pole to another in warmer climates, reaching a dip of about one foot at the center. Several jointing and splicing techniques are available, as shown in the diagrams.

Underground Conduit

An underground conduit instead of poles is generally planned for the installation of urban lines. The most common method is to

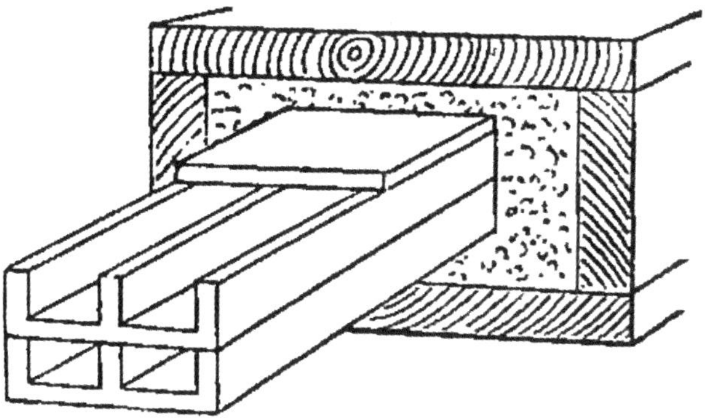

Rectangular conduits in terracotta. William J. Hopkins, *Telephone Lines and Their Properties* (New York: Longmans, Green, and Co., 1899), 19.

install rigid sleeves with openings within easy reach, to ensure the maintenance and repair of wires. These sheaths run in straight lines between each opening. Sharp angles are avoided as much as possible to prevent abrasion of the cables. There are three types of conduits: plant material, iron, or concrete. They must be rigorously separated from gas and water lines to avoid mold, the degradation of the cables, and even explosions within the network. Early on, ropes provided access to all the cables in a sheath, but this arrangement was quickly replaced by a system of short shafts installed at each maintenance opening that hold the cables together.

Distribution

Distribution of the wires in the conduits to each subscriber is done by short overhead lines from a distribution pole. This pole is usually located in the nearest central position relative to a group of subscribers.

The cables run along the top of the pole through a copper wire that connects to the cable box, which protects them against lightning strikes and mold. This box surrounds the pole and is about thirty inches in diameter. A circular platform with a railing and a trap door allows the lineman to gain access to the lines. Iron steps on the poles enable the lineman to reach the higher lines. The meeting between the cable and wires occurs inside the box as illustrated by the diagram.

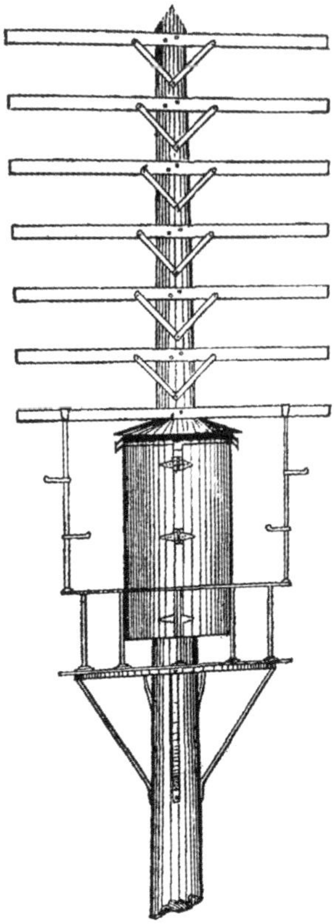

Distribution poles. William J. Hopkins, *Telephone Lines and Their Properties* (New York: Longmans, Green, and Co., 1899), 9.

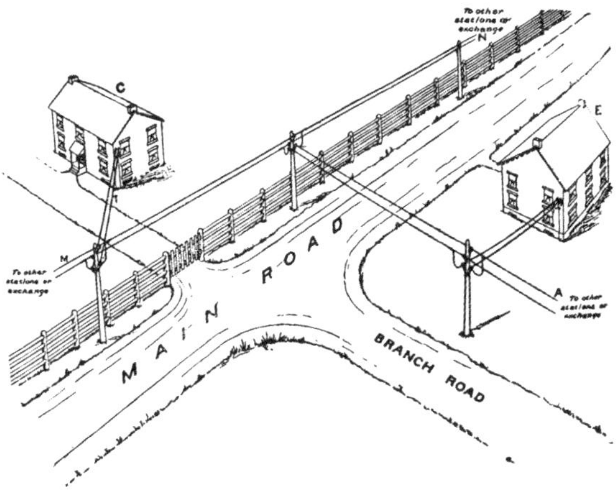

Distribution poles. William J. Hopkins, *Telephone Lines and Their Properties* (New York: Longmans, Green, and Co., 1899), 9.

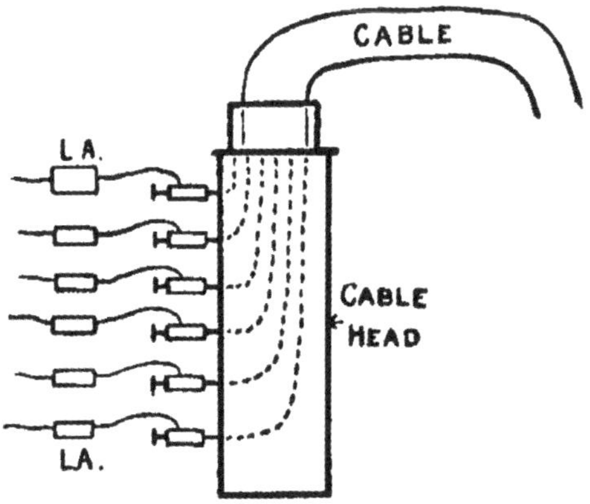

Connection of cables and wires. William J. Hopkins, *Telephone Lines and Their Properties* (New York: Longmans, Green, and Co., 1899), 30.

Cable and Insulation

The principle of the cable is to group together a series of insulated wires covered with metal or other materials. To avoid induction, the two wires of the same circuit are twisted together. Fibrous materials such as cotton, jute, paper, or rubber are used for the insulation of the cables.

Long Distance

The construction guidelines for long-distance lines set the highest levels of standards and requirements. The first long-distance line, between Boston and New York, was established in 1884. It was planned with particular attention to the quality of its infrastructure. Live Canadian cedar poles were selected and treated with paint. The posts measure around six inches in diameter at the top and are slightly thicker at the base, with both ends cut into squares. They are at least thirty-five feet high, with a capacity of four arms, each carrying ten wires. The joint of each arm is cut in advance, so these can be installed later if necessary. Some of the poles installed by the American Telephone & Telegraph Co. would reach heights between seventy and one hundred feet, to avoid interfering with other activities, especially in town. The arms, always made of pine, with a section of 4¼ by 3¼ inches, are placed twenty-four inches

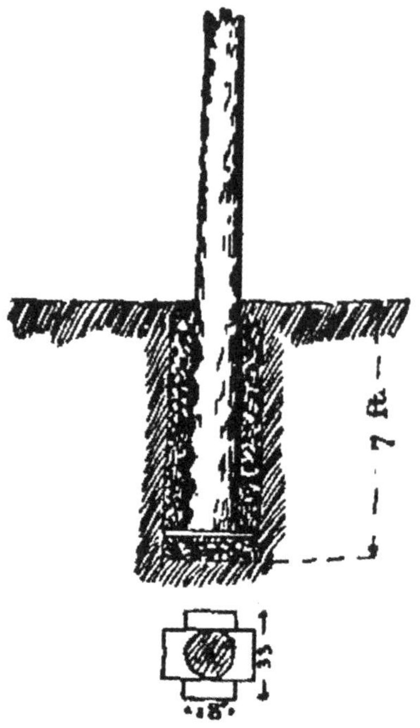

Pole set in the ground two meters deep. William J. Hopkins, *Telephone Lines and Their Properties* (New York: Longmans, Green, and Co., 1899), 42.

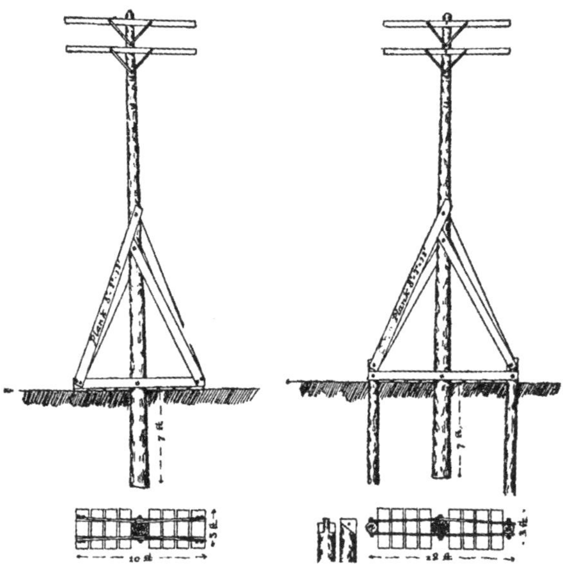

Pole support system. William J. Hopkins, *Telephone Lines and Their Properties* (New York: Longmans, Green, and Co., 1899), 43.

apart on centers, the upper arm being ten inches from the top of the pole. The post is dug into the ground to a depth of six feet: a hole is drilled first, then the post is inserted, and the hole is filled with one part cement, two parts sand, and five parts broken stones. A thin layer of mortar is then poured over the hole. When the soil is very bad, a system of bolted arms, reinforced with piles, is constructed.

On straight lines, the posts are spaced about one hundred thirty feet apart, with the arms alternately extending face-to-face or back-to-back. When the distance between two poles reaches two hundred feet, the arm is placed in such a way as to push toward the post. At the end of the line and for curves, the arms are always placed toward the outside of the pole as well. Lines crossing the territory must be treated with particular care compared to urban lines or lines running along railroads, while respecting the same principles. Long-distance lines should follow the topography of the ground, with the top of the poles at the same height.

In curves, the wire must be taut, and sloping is to be avoided. Cedar poles do not split in half, but they tend to break into small pieces, either at the base of the pole or under the crossing of the arms. Several bracing techniques are used to reinforce the poles, depending on the situation, as shown in the diagrams.

Technical descriptions for power transmission complete Hopkins's manual on line planning. In this pioneering handbook, the infrastructure is associated with a form of aesthetic pleasure, as we have seen, made explicit in the text by the association between the good functioning of the line and its beautiful appearance. This consideration of the materiality of infrastructure makes it possible to judge not only its technical capacities, but also its real presence in the territory, and is crucial for interpreting the subsequent development of communication networks. As soon as we stop thinking of infrastructure as a system of objects and associate it with immaterial forms of transmission and invisibility, we overlook the importance of its actual mechanical functioning.

"Nothing screams more loudly of the still-developing-nation status of the United States than the creosote-treated poles, all slightly out of perpendicular, marching along almost every road as they once marched across the plains in hoofprints of the Pony Express."[8] John Stilgoe long afterward recognized the telegraph and telephone infrastructure as an expression of the ongoing development of the "New World." He also emphasized the ways

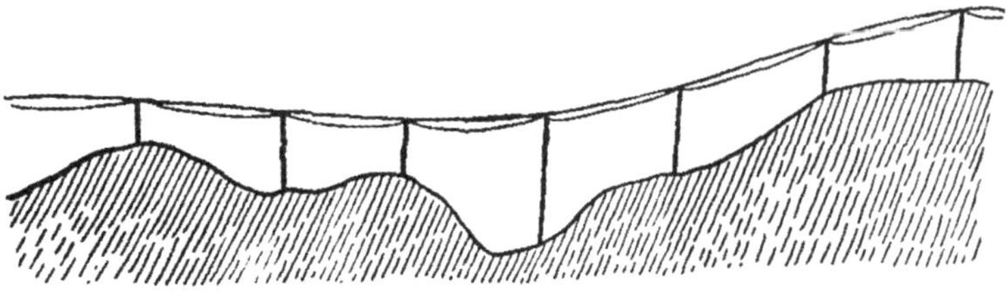

Post heights adjusted to topography. William J. Hopkins, *Telephone Lines and Their Properties* (New York: Longmans, Green, and Co., 1899), 44.

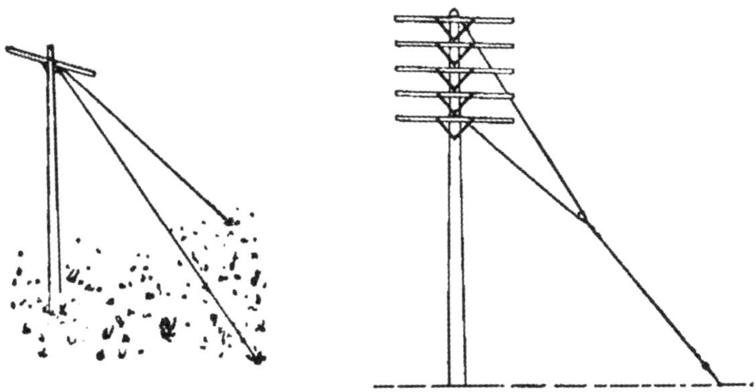

Good height techniques. William J. Hopkins, *Telephone Lines and Their Properties* (New York: Longmans, Green, and Co., 1899), 48.

Bad height techniques. William J. Hopkins, *Telephone Lines and Their Properties* (New York: Longmans, Green, and Co., 1899), 49.

in which these networks remain essential and handcrafted, an orthogonal grid of wooden poles and copper wires that proliferated from 1840 on, alongside other infrastructures such as the road and the railroad. While other nations adopted different techniques and materials over time, such as underground cables or steel pylons, this was not the case in the United States, "land of cheap timber, vast distances and an easygoing willingness to accept the poles that warp and twist and finally rot."[9] The simplicity and rawness of these infrastructures would end up representing an attitude, a mentality, and most significantly, a culture. Since their installation, thin and flexible wires enable the deployment of communications on an immeasurable territory by tracing lines, creating superimposed horizons covering the rural and urban landscape of the country. In this manner, they celebrate a union between the technological empiricism and romantic pastoralism of American culture.

Today, this simplicity, consistency, and durability of an infrastructure that continues to rely on wood and copper still embodies a certain originary spirit of progress, while presenting major complications from the perspective of use. These antiquated infrastructures cause some of the longest and most frequent power outages in "developed" countries, with important consequences for security and the economy. These problems are due not only to strong meteorological events, but also to the normal course of nature, vegetation, and animals, especially squirrels. Several political, economic, and cultural forces lead to a form of inertia in the face of possibilities for change—a renewal that is nevertheless essential in the face of current ecological imperatives—of "the grid," the electrical network of the American continent.[10] Furthermore, going beyond the pure analysis of their workings, the linearity of telecommunications infrastructures and their materiality afford different forms of reading.[11] The physical permanence and structural simplicity of these objects still recall the legendary times of their installation and their extension throughout the territory, the times of conquest of the West.

To this day, the line functions as an object that interrogates the landscape and attracts the gaze of multiple observers by its physical presence, its iconic silhouette, and the way it spreads out across the territory. The linear infrastructure described here enables communication at distance between individuals. It takes the place of a moving network of human beings—or sometimes animals—that once crossed the territory. Having become fixed in place, the telecommunications network allows faster exchanges between individuals, through the movement of electricity and the transmission of sound signals. This complex relation between a static network and the resulting movement of messages and people has been described by the great Canadian media theorist Marshall McLuhan, who saw a new model of decentralization in the electrical network, like pre-literate, or "tribal" societies.[12] Thus, if linear writing forged Western culture by a hegemony of the visual perception, the specialization of individuals, and the establishment of a sedentary society, global and instantaneous communication would again transform the experience of space through a new state of nomadism, of decentralized groups that once again perceive the primacy of "acoustic space," of tactile over optical perception.

This wild and primitivist conception of the mechanized territory of North America maintains the myth of a nature to be conquered. The category of the landscape as a framework to be observed from a unique point of view is thus overcome, in favor of the idea of an unlimited and multifocal space that is lived through a series of encounters and relationships.[13] In this wild space of the North American territory, the encounter, even the frictions with technical objects—the infrastructures that range across the topography of the Earth—along with their texture and their sonority, is often characterized by clearly present and accentuated materiality.

2 *Aesthetics: The Sublime and the Ugly*

The physical objects of telecommunications infrastructures function as signs of technical progress, as instruments for new forms of social life, as models for conceiving the modern city, and as visionary objects for the world of the future. They play a significant role in various forms of artistic representation. Their diffuse presence, their imposing or filigree forms, their systematic proliferation in the landscape, and their allegorical value have inspired sublime evocations of the landscape and symbolic interpretations of their presence. The reception of these objects placed on American soil reveals different aspects of the country's culture in the face of new technological artifacts and emblems of progress. In *American Technological Sublime*, David Nye articulates the difference between what he defines as the geometric sublime, which represents static and monumental infrastructure capable of dominating the natural landscape, and the dynamic sublime—embodied in the infrastructures of the railroad, the telegraph, and the steamboat—that emphasize the movement of information and the conquest of time and space.[1] This apparent dichotomy characterizes the modern territorial transformation, implemented through the construction of forms and material grids. In this vision, as we shall see, the aesthetics of the line illustrates a kind of founding ideology of the nation and the culture of the North American pioneers. This connection is founded in history and provides a fertile ground for artistic readings that sublimate the technical materiality of networks.

While this study is mainly focused on the material, aesthetic, and symbolic value of the lines, it also repeatedly tries to detect the sound element evoked by representations of the technologies of signal transmission, in other words the sound networks. Following the materiality of the wire generally leads to intermittent reappearances of the phenomenon of sound in various forms: from the poetic evocation of musical notation (in the lines) to new practices of modern telecommunication—penetrating the architecture through networks—along with the pseudo-musical harmonies emitted by the electric wires that cross the landscape. Indeed, despite the impossibility of visualizing sound, the increased capacity for propagating the signal of modern telecommunication has, in a more or less direct way, inspired several aesthetic representations of these networks.

History shows various examples of attempts to measure the atmosphere or a medium by sound.[2] A wet or dry, a dense or diffuse environment sounds differently, which means that each environment translates its own particular material qualities into a sound signal. The line functions like other "elemental" objects, like great musical instruments that, thanks to the transmission of the atmospheric medium—"the vibrating force of things"—sound and make explicit the properties of the air.[3]

The "telegraph harp" was a source of fascination for the American poet Henry Thoreau, who in the mid-nineteenth century described the harmonious crackling of the landscape's open horizons. In his diary, Thoreau evoked images of the resonating wood, the wires vibrating in the wind, an electric tree in the middle of a forest, a telegraph harp that sounds loud in the rain, music that is close to the divine—that is like a bright color, red, blue, or green, while everything around is black and white—to express a feeling of communion not only with nature but also with the human enterprise, both being part of a common work of creation.[4] The opposition between the natural and the artificial world in this North American cultural context is not perceived as an antagonistic relationship as it is more commonly considered in Europe.[5] The rapprochement with nature, evoked by the transcendentalist poet, must perforce include the great works and infrastructures that convey the sense of the territory. Their sound becomes a prodigious sign of progress, but also a trace of the divine.

From Burke to Kant, the sublime is defined as an emotional condition of the observer as an individual, who projects his own interior state out into the external world. The experience of the sublime draws attention away from the physical object to a mental and emotional state of perception. These ideas landed in North America during the Jacksonian period, as part of the development of a progressive culture that characterized American society at the time. If the association between the sublime and technology may seem at first glance a form of antinomy, with the first term being born of a confrontation with nature in its wildest state and the second the artificial product of human ingenuity, it is precisely by bringing together these two originally unrelated words that a new semantic designation appears. The latter reflects a new condition of the landscape that emerged in the North American context more and more clearly over the course of the nineteenth century, when the association between the existing territory and the extension of infrastructure became a primary condition for taming the territory—for both historical and geographical reasons.[6]

As Nye observes, the North American experience of the sublime is part of a popular and collective history. It finds expression in enthusiasm for new forms of local tourism—for landscapes that are both natural and technological—for majestic infrastructures such as dykes, bridges, and electrical stations.[7] According to Susan Glickman in her study of the Canadian context in the late nineteenth century, the experience of the sublime resembles "a new kind of religion" born of the confrontation between the grandeur of nature and the great human endeavors that spread out over those vast territories.[8] For example, a 1926 photograph by John Vanderpant shows two infrastructures that are emblematic for that country, both for their economic value as signs of development and their iconic status through characteristic forms that mark out the territory: in the background, the gray and imposing mass of a grain silo, and in the foreground, against the light, an electric pole in all its lightness, along with a human figure so small in proportion. Located in Vancouver, in the province of British Columbia, this composition along with its blurry and chiaroscuro effect perfectly fits to a sublime account.

By examining two artistic representations, this chapter affords an interpretation of the modern imaginaries in the face of the installation and proliferation of telecommunications infrastructures. A comparison of two images—*American Progress* and

Tina Modotti, *Untitled (telephone wires)*, 1923–1930. Courtesy of Isabel Carbajal Bolandi. Museum of Modern Art (MoMA), New York, USA. © 2025. Digital image, The Museum of Modern Art, New York / Scala, Florence.

Through the Rocky Mountains, a Pass on the Canadian Highway—by the American painter and lithographer George Gast in 1872, and Canadian Lucius Richard O'Brien in 1887, respectively, highlights two different ways of celebrating the electrical infrastructure embedded in the vast landscapes of the North American territory.

The first image, Gast's *American Progress*, is a hymn to the ideal of the nineteenth century frontier.[9] The ensemble of transportation lines—the horses, the Pony Express, the caravan, the railroads, and the naval traffic—expresses a kind of temporal and technological

Photograph by John Vanderpant, Vancouver, Canada, 1926. Canadian Centre for Architecture, PH1987:0979.

ascension reflecting the inevitable race of progress. Similarly, the sequence of figures depicted marks a transition away from the indigenous people to the Euro-American pioneers who settled these territories. The combination of linear objects and moving figures illustrates both the evolution of time and the progress of infrastructure. A large woman dominates the center of the image. Dressed in a flowing robe, she flies over the landscape, moving in a westerly direction. With the star of the empire on her forehead, she carries the "common schoolbook," an emblem of national education.[10]

John Gast, *American Progress*, 1872. Autry Museum of the American West, Los Angeles.
Autry Museum 92.126.1

OVERLAND MAIL
US

Her left hand unfurls an electric wire, emphasizing how this other object of progress physically imposes itself on the expanse of the territory. The recurrent association of female characters with electrical invention is not simply anecdotal and has been the subject of several studies.[11] Electricity is a woman. In the rhetoric of the time, she has her qualities and her defects. Capricious, she has the charm of the invisible force, she contains the secrets of nature, she has the magical powers of a fairy, as well as the occult power of a witch.[12]

The image of the thread accentuates the linear sense of movement by merging with the American pastoral landscape in the background. This combination of nature and technology allows us to think of the frontier as an expanding line across the country, embodied in the myth of the conquest of the West, and materialized in the migration of people along with the construction of an infrastructure that facilitates national unity. The photographic documents preserved in the AT&T archives celebrate the work of setting up telecommunications networks: the digging operations, the underwater immersion, the aerial laying of wires and cables by the so-called linemen, who became true heroic figures.

The portraits of these men with their work tools, sometimes in extreme cold and bad weather conditions, seem to be designed to highlight the physical and manual effort of the human enterprise. Sometimes they pose in very haphazard positions, in groups or alone, on thin, high poles, standing out, occasionally against the light, from the rural or urban landscape of the American territory.

In his celebratory speech for Independence Day in 1851, Secretary of State Daniel Webster proclaimed, "The net-work of railroads and telegraph lines by which this vast country is reticulated have not only developed its resources, but united emphatically, in metallic bands, all parts of the Union."[13] The metallic band is no longer a mere technical object, but a true symbol that evokes a national feeling.

The second image, O'Brien's *Through the Rocky Mountains, a Pass on the Canadian Highway*, shows a wild landscape built on different layers: a turbulent river, a deep forest, snow-capped mountains, and a bleak sky. The accentuated verticality of the format and composition, emphasized by very slender trees and mountain peaks, is broken by a single horizontal line: a moving train, a railroad, a bridge, and a series of electric poles. The painting is part of a commission made by William Cornelius Van Horne, one of the main promoters behind the construction of the Canadian Pacific

Lucius O'Brien, *Through the Rocky Mountains, a Pass on the Canadian Highway*, 1887.
Private collection, Toronto. Cowley Abbott Fine Art.

Railway (CPR).[14] This landscape is located precisely in the province of British Columbia: the train crosses the bridge over the Kicking Horse River, climbs Big Hill, and on the left is Mount Stephen. In the foreground below, a broken tree trunk and white water. This scene of natural sublime accompanies the sense of grandeur associated with the construction of the CPR, which opened up a road from East to West, across a country that was both immense and untamed. More realistic than Gast's *American Progress*, O'Brien's painting also associates the transportation infrastructure with the structures of electrical transmission, through a string of utility poles, both essential to nation building. This single horizontal line in the middle of a painting built on a pronounced verticality reinscribes in a different way the idea of linear imperialism, the territorial conquest toward the British Columbia Coast. While in the United States the construction of the country could be said to have been relatively symmetrical, moving the borderline from coast to coast, but also expanding from North to South during the Revolutionary War, the axis of Canadian colonial expansion was more radically unidirectional, following the course of the St. Lawrence River, across the Great Lakes, across the prairies, and down to the Pacific Coast.[15] Harold Innis emphasizes how Canada was built on commercial and administrative imperatives from Europe, imposed on the local culture and landscape.[16] The transportation of raw materials from Canada was a major factor in the development of the country. They were transported to the East Coast ports, then to England and on to the continent, again in a linear physical movement, but this time in the opposite direction from the movement by which the country was formed. The notion of "linear imperialism" was taken up explicitly by Barbara Novak to describe the physical and political rush that characterized the American landscape of the nineteenth century.[17] The machine age developed in parallel with the age of empires, and the technical infrastructure was thus associated with the means of control put in place by Europe over societies it deemed archaic.[18]

North American telecommunications propaganda contributed to the creation of an aesthetic of the line, emphasizing the simple and powerful structure that determines its infrastructure. This motif embodied the principle of both modernization and, as we have seen, national cohesion. Two central figures within the American Bell Telephone Company and its Canadian division, director Theodore Vail and entrepreneur Charles Fleetford Sise, shared

Plant construction, cable laying, Michigan, May 1917. Courtesy of AT&T Archives and History Center.

Michigan State Telephone Company, laying underground cable, May 1917. Courtesy of AT&T Archives and History Center.

The Mackinaw submarine conducts a cable attached to a raft to shore, September 1918. Courtesy of AT&T Archives and History Center.

Cables laid heavily in the sea, September 1918. Courtesy of AT&T Archives and History Center.

Cables on a boat, Michigan State Telephone Company, September 1918. Courtesy of AT&T Archives and History Center.

Unrolling a cable and guiding it underground, Western Electric, ca. 1900. Courtesy of AT&T Archives and History Center.

Construction crew digging a trench to lay cable, Pittsburgh, 1910. Courtesy of AT&T Archives and History Center.

Laying cable between Boston and Washington, DC, 1910. Courtesy of AT&T Archives and History Center.

Concrete roller between Laurel and Contee, Maryland, December 1910. Courtesy of AT&T Archives and History Center.

Plowing trench, Boston-Washington, DC cable, November 1910. Courtesy of AT&T Archives and History Center.

Plowing trench between Lauren and Contee, Maryland, Boston-Washington, DC cable, December 1910. Courtesy of AT&T Archives and History Center.

Assembling concrete splicing box, Boston-Washington, DC cable, November 1910. Courtesy of AT&T Archives and History Center.

Scraper used for backfilling trench, Boston-Washington, DC cable, November 1910. Courtesy of AT&T Archives and History Center.

Trenching machine, Northeast, North East, Maryland, Boston-Washington, DC cable, 1910. Courtesy of AT&T Archives and History Center.

Portrait of linemen with their working instruments, 1910. Courtesy of AT&T Archives and History Center.

Line installation team, 1900. Courtesy of AT&T Archives and History Center.

Line installation team, 1930, New York Buffalo Line, Little Falls, New York. Courtesy of AT&T Archives and History Center.

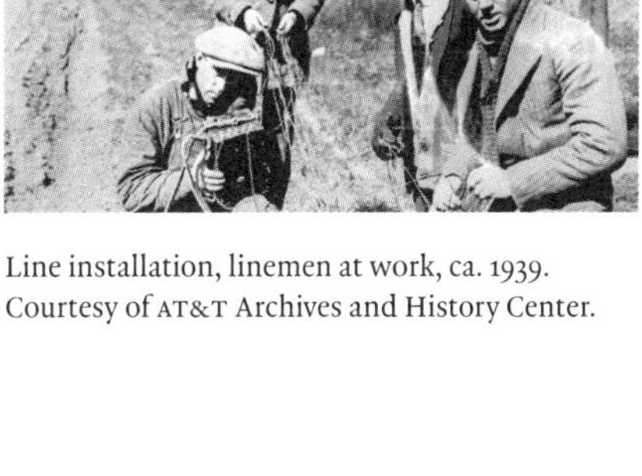

Line installation, linemen at work, ca. 1939. Courtesy of AT&T Archives and History Center.

Pole installation, New York Telephone Company, September 1897. Courtesy of AT&T Archives and History Center.

Linemen of Southwestern Bell Telephone Company, 1930. Courtesy of AT&T Archives and History Center.

Line installation, lineman at work, 1940. Courtesy of AT&T Archives and History Center.

Emergency restoration, n.d. Courtesy of AT&T Archives and History Center.

Linemen, line repairers, n.d. Courtesy of AT&T Archives and History Center.

New York Telephone Company linemen at work, 1958. Courtesy of AT&T Archives and History Center.

Lineman installing telephone lines, 1970. Courtesy of AT&T Archives and History Center.

Aerial linemen in Albany, Georgia, 1905. Courtesy of AT&T Archives and History Center.

Linemen posing on a pole, 1920s. Courtesy of AT&T Archives and History Center.

Linemen posing on a pole, n.d. Courtesy of AT&T Archives and History Center.

Installation of cable between two buildings, n.d. Courtesy of AT&T Archives and History Center.

Aerial lineman, Indiana Bell, 1932. Courtesy of AT&T Archives and History Center.

Linemen on a platform to test cable, 1911, Monongahela City, Pennsylvania. Courtesy of AT&T Archives and History Center.

Linemen team, ca. 1900. Courtesy of AT&T Archives and History Center.

Lineman on the top of a pole adjusting a joint, 1945. Courtesy of AT&T Archives and History Center.

Lineman on a distribution pole, n.d. Courtesy of AT&T Archives and History Center.

Two linemen sitting on telephone cables, n.d. Courtesy of AT&T Archives and History Center.

Lineman demonstrates how to grip a pole for the film shooting of *Daughters of the Night*, ca. 1924. Courtesy of AT&T Archives and History Center.

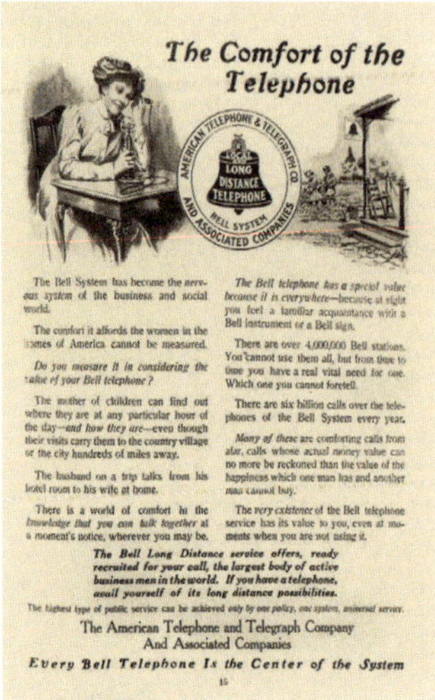

AT&T advertising, 1909. Courtesy of AT&T Archives and History Center.

AT&T advertising, 1938. Courtesy of AT&T Archives and History Center.

Postcard from 1908. Special Collections, Lancaster University Library.

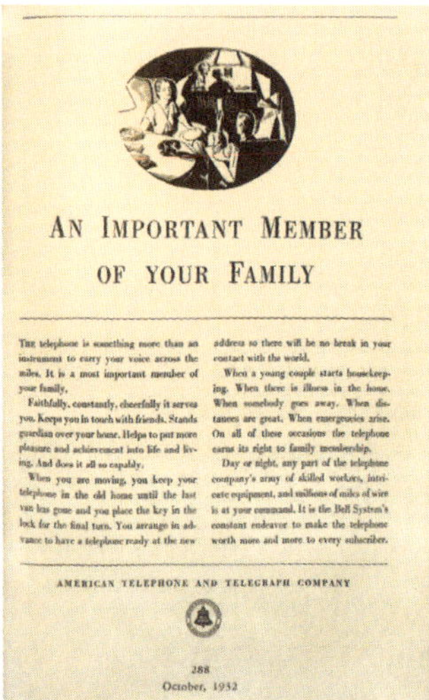

AT&T advertising, 1932. Courtesy of AT&T Archives and History Center.

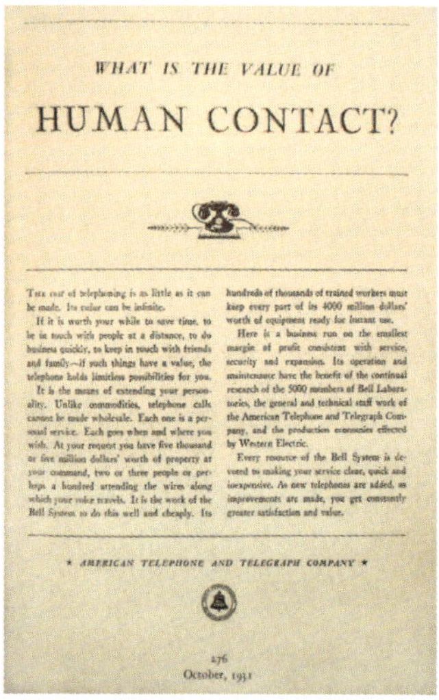

AT&T advertising, 1931. Courtesy of AT&T Archives and History Center.

AT&T advertising, 1932. Courtesy of AT&T Archives and History Center.

AT&T advertising, 1933. Courtesy of AT&T Archives and History Center.

AT&T advertising, 1934. Courtesy of AT&T Archives and History Center.

AT&T advertising, 1940. Courtesy of AT&T Archives and History Center.

a vision: to build a national network for the United States and Canada by merging local companies into a single long-distance infrastructure system.[19] Although this process of modernization of the country is generally described in terms of its favorable reception and as an element of national cohesion, one must qualify the way in which this technological turn was perceived in reality and note mixed feelings of both enthusiasm and suspicion. During the early part of the so-called Progressive Era, from 1890 to 1920, a number of people expressed their anxiety about this complexification of society through a series of demonstrations, protests, and other types of resistance.[20] While still emphasizing the scope of the technological revolution, advertising campaigns went beyond the simple ideology of progress to associate the new infrastructures imposed by modernity with the reassuring values of comfort, tranquility, family unity and security. The vortex of progress was constantly blurred by visions of peaceful daily life.

Many interpretations of telecommunications networks focus on the physical constitution of the objects studied and attempt a simplified reading of their forms, sometimes to the point of geometric reduction. The figure of the line, whether flexible or rather straight, appears repeatedly. First, the horizontal lines—the electric wires—which double or cross the existing horizon, where earth

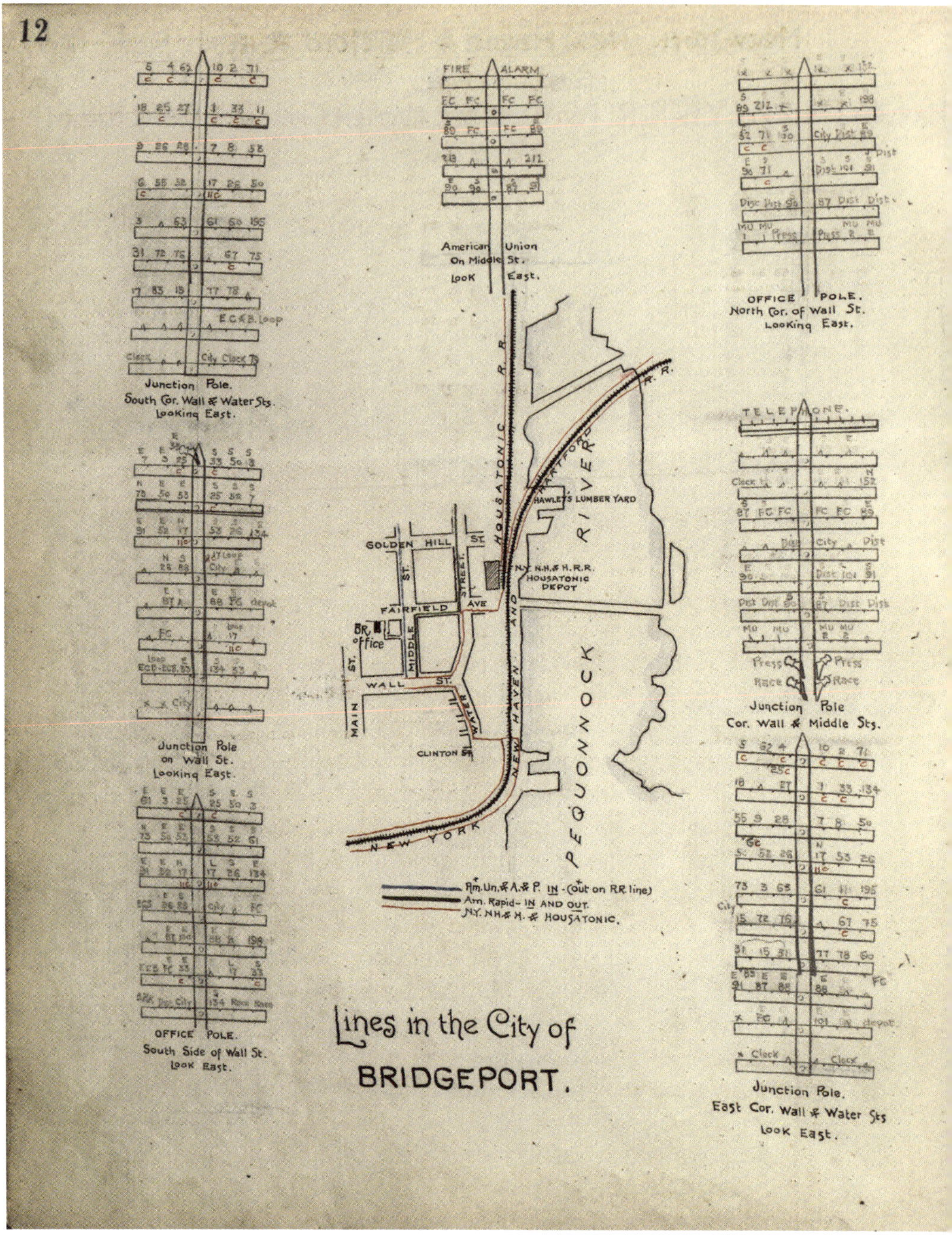

Directory of lines, poles, and arms, New York, New Haven & Hartford R.R., 1904. Western Union Telegraph Company Records, Archives Center, National Museum of American History, Smithsonian Institution.

New York, New Haven & Hartford R.R.

Main Line.

rected to
ly 1st 1904

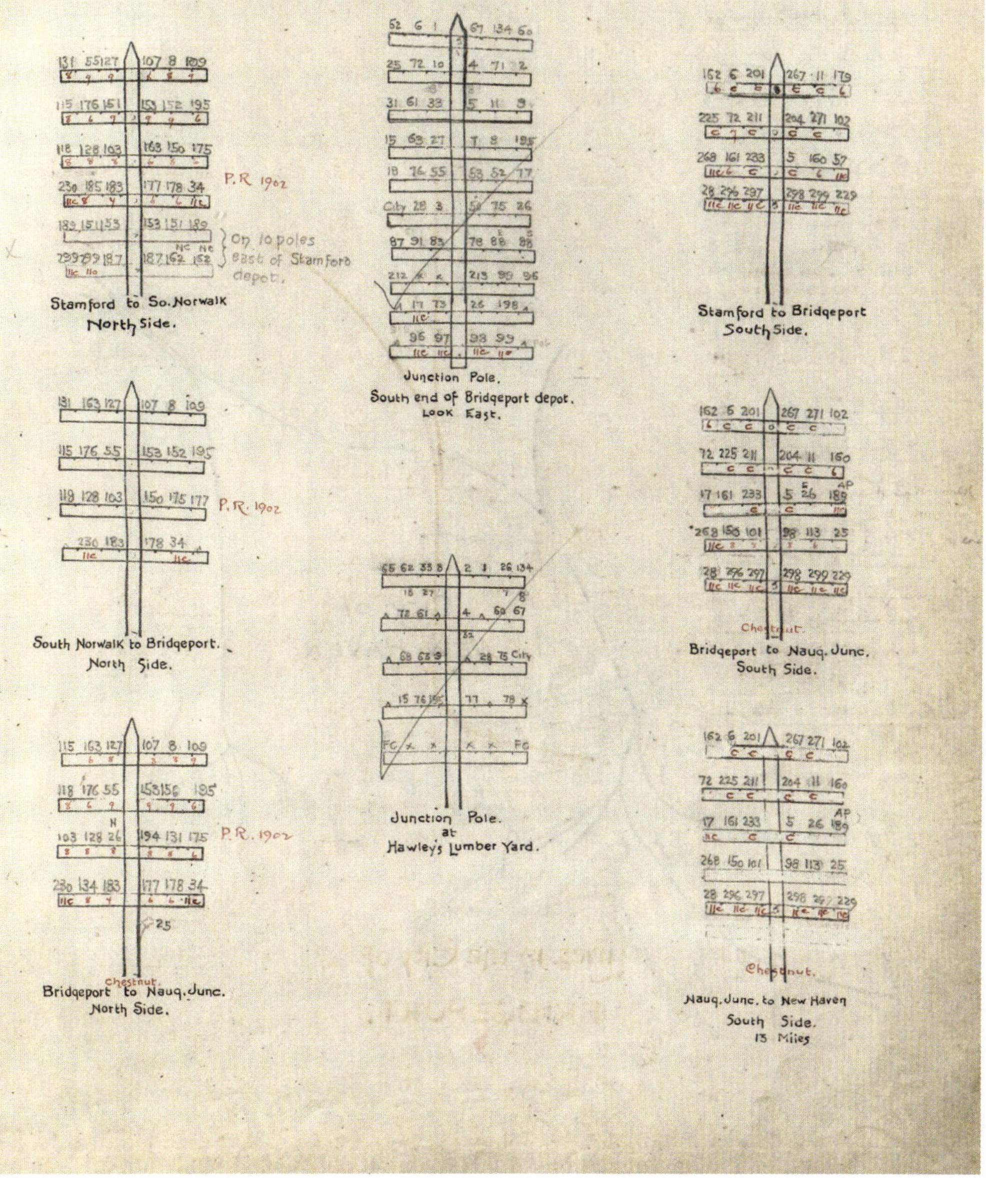

AT&T advertising, 1912. Courtesy of AT&T Archives and History Center.

AT&T advertising, 1914. Courtesy of AT&T Archives and History Center.

AT&T advertising, 1921. Courtesy of AT&T Archives and History Center.

and sky meet. And then the vertical lines—the electric poles, the pylons—sunk into the ground and supporting the wires, mark a form of technological human conquest of the territory. So do the crossed lines of the system and the network, represented on plans, sections, axonometries that illustrate the performative character of the architecture, the conduits of circulation of the flows, their functioning.

But it is precisely this demand for simplification, reduction and schematization that needs to be explored and that can reveal a lot of information about the status of these objects, about their relationship with space and the human being, as well as about the way we observe them. Telecommunications networks, electric wires, become lines in certain types of representation to symbolize their power of connectivity, their occupation of the territory on a large scale, their agility and structural simplicity. But they are also reduced to lines in order to speak the language of planning, to enter the thinking and culture of architecture and urban engineering, which cannot do without these modern infrastructures and their constant presence in the territory.

In the conclusion of his book *Lines: A Brief History*, the anthropologist Tim Ingold emphasizes the intimate link between the line and the atmosphere, because lines are for him the result of an action and are a sign of life. The atmospheric environment is thus

A line of broken poles after a storm in Hawthorne, Illinois, February 20, 1913. Courtesy of AT&T Archives and History Center.

The result of cyclone and flood, Terre Haute, Indiana, May 1913. Courtesy of AT&T Archives and History Center.

Street in front of Cincinnati, Ohio, house during a flood, May 1913. Courtesy of AT&T Archives and History Center.

Wires remain intact after storm in Columbus, Ohio, May 1913. Courtesy of AT&T Archives and History Center.

Boatload of wires during a flood in Pittsburgh, Pennsylvania, May 1913. Courtesy of AT&T Archives and History Center.

Loading a shipment of emergency cables in Pittsburgh, Pennsylvania, May 1913. Courtesy of AT&T Archives and History Center.

Broken poles held by a transmission wire in Columbus, Ohio, May 1913. Courtesy of AT&T Archives and History Center.

A typical street in New York before the wires were buried in cables underground. Photo taken from the pamphlet *The Biography of the Telephone*, p. 13, published in 1913. Courtesy of AT&T Archives and History Center.

Telephone wires in New York, 1887. Everett Collection, Inc.

the "site where these lines of life intertwine,"[21] not only as signs, but also explicitly as material objects. Lines materially cross the space with all its thicknesses, its textures, and its sonorities. They maintain a close relationship with architecture and the territory: they react to the climate and bad weather; they undergo or cause breakdowns of the infrastructure. If modern inventions necessarily introduce the possibility of accidents, likewise the wired infrastructure of telecommunications has been the cause of numerous technical and environmental disasters.[22]

Replacement of elevated wires with underground cable in Sixth Avenue, New York, 1882. *Frank Leslie's Illustrated Newspaper* (January 29, 1887). Courtesy of AT&T Archives and History Center.

In fact, it was not long before cable congestion in dense urban areas became a problem, even a potential source of danger, as illustrated in the case of New York. In the early 1880s, on some of the city's streets, the poles reached a height of about one hundred feet, with thirty arms, each carrying about three hundred electric wires.

The urban landscape was obstructed by a disordered system of wires that caused difficulties. In particular, communication was frequently interrupted during the winter, for days or even weeks because of snowstorms that broke down the infrastructure. For these reasons, a New York State law required all cities with more than five hundred thousand inhabitants to install cables and wires underground by November 1, 1885.[23] However, the technology was not ready. The underground wires could not yet operate for more than a couple of miles, and the law was therefore repealed.[24] It was only after 1890, thanks to the initiative of Theodore Vail, president of the firm, that AT&T began to produce insulated cables using cardboard crumpled with air—instead of wax, resin, and

New York's New Street during the Great Blizzard of 1888. Telephone service was paralyzed from March to May. The Museum of the City of New York. Granger—Historical Picture Archive / Alamy Stock Photo.

oil—capable of running through a city like New York. This solution came just after the greatest catastrophe in the history of the American telephone system, the Great Blizzard of 1888.

The rapid densification of these networks during the period of development of industrial cities at the end of the century was therefore more than just a source of progress and comfort. Although today most large metropoles use underground networks for the placement of wires and cables, there are still cities such as Tokyo where the concentration of overhead wires remains imposing: only about 8 percent of the network is underground. This mass of wires that characterizes the local urban landscape actually constitutes a real danger in the face of the country's seismic threat, which also makes it difficult to bury the wires and makes them an important subject of political propaganda.[25]

The objects of modern telecommunications have inhabited other types of landscapes and imaginaries in other contexts. In Europe in particular, the artistic representation of these objects is less tied to national foundational myths. In several representations from the beginning of the twentieth century, these technical objects occupy the foreground; nature fades away to make room for these different signs, accentuating a form of radical opposition and exclusive vision, between the natural territory and the technological enterprise. Very often, the electrical infrastructure is then associated with the urban landscape. It inhabits the great metropolitan centers and joins the products of modernity. If the sublime North American illustrations insist on the confrontation of modern infrastructures with the grandeur of nature, other images describe the installation of electrical networks connected to the city, often in a kind of univocal relationship, of confidence or distrust, toward progress that erases all traces of the rural world as well as the natural landscape.

The photographs reproducing Mies van der Rohe's celebrated photomontages for the 1921 Friedrichstrasse skyscraper competition in Berlin show a view taken from the north side of the project, which maintains in the foreground a line of poles supporting electric wires. This line, which emphasizes the depth and perspective of the frame, stands out strongly against the background of the glass building, which is represented as a transparent mass, a silhouette held together by a more assertive orthogonal structure. Although the electrical infrastructure, such as the street lights, the hydrant, and the few human silhouettes, would disappear in the final version of the drawing, in favor of a more refined and contrasted rendering of this urban scene, it is significant to note how versions of the photomontage have been widely circulated in publications and exhibitions, constituting not merely provisional stages of the drawing, but documents in their own right.[26] The final drawing produced for the competition, based on the photographic images, becomes a plastic image emphasizing the black mass of the existing buildings over the more transparent and light mass of the glass skyscraper—adding a bridge as a vanishing point

Ludwig Mies van der Rohe, photomontage for the Friedrichstrasse skyscraper competition, Berlin, 1921. Museum of Modern Art, New York. Ullstein Bild / Granger Collection © 2023, ProLitteris, Zurich.

in the background—while the photomontages reduce the degree of abstraction of the image to a more complex and layered perception of the city.

These documents, like the drawing preserved in the collections of the Museum of Modern Art in New York, would circulate widely, generating several interpretations regarding the architect's work

Ludwig Mies van der Rohe, perspective view for the Friedrichstrasse skyscraper competition, Berlin, 1921. Museum of Modern Art, New York. akg-images © 2023, ProLitteris, Zurich.

process and the status of these same images.[27] They undoubtedly testify to a fascination for the technical objects that delineate the modern city, while making them disappear when the plasticity of the architectural object is brought to the fore.

In his idealized project for the American suburb called Broadacre City (1932–1958), to which we will return in more detail later, Frank Lloyd Wright also seems to have evacuated the technical materiality of the networks, while developing a real preoccupation with the aesthetic implications of the technological regime of his plan. While he celebrates the magnitude of transportation infrastructure, he also clearly despises the modern landscape of wires and cables. Relying on the advancement of wireless technologies and the future development of underground networks, he believed that "all the poles and overhead wires will be a bad memory of ugliness and danger." In *Living City*, his compilation of essays on Broadacre City, Wright emphasized the need to bury those "ugly scaffolds," in other words the vertical grid of poles and power lines that occupied the skyline of the American industrialized territory. He writes: "This imposition of crude, utilitarian scaffolding, incurred during the infancy of our haphazard growth as a mechanized nation, invariably does violence to our own character as well as the countryside, violates to this day all the finer sensibilities of American life. Such crudity disappears in Usonia."[28]

For Wright, the memory of the origins of modern urbanism and its remnants was to make way for an invisible, even subterranean infrastructure. The visibility of telephone poles and wires made evident their economic and social values attached to the first period of concentrated capitalism: "Evidences of these crude if fundamental ways of money-making our pioneering days accepted are seen no more. Obtrusive, offensive scaffolding of any kind will be taken down so that the culture of our civilization may now appear, though as late as the middle of the twentieth century."[29] Looking at the architect's views of Broadacre City, one recognizes different structures distributed within the one-acre grid that organizes the land. However, there are no poles or wires crossing the horizon of this city. The sky is punctuated only by the heights of some tall buildings and futuristic personal helicopters, in other words points rather than lines.

In fact, even today, the grid of poles and overhead wires is far from being a simple archaeological relic, especially in rural areas. It continues to connect us to the origins of the industrial age and its period of infrastructural expansion. Like any sign of progress, this embodiment of the modern spirit is occasionally confronted with other expressions of absolute rejection linking the technical objects of telecommunications to the period of unbridled industrialization and capitalism. The progressive enthusiasm is on occasion accompanied by forms of resistance which are reflected in

Robin Boyd, *The Australian Ugliness*, cover, 1960. © Robin Boyd, courtesy Robin Boyd Foundation.

the cultural debate of the time. Modernity leads not only to forms of exaltation among technicists, but also to feelings of rural and backward-looking nostalgia that are translated into other types of discourse and imaginary evocations.

From the 1950s onward, the Townscape movement reactivated a taste for the preindustrial city through the English magazine *The Architectural Review* and was subsequently spread through other attacks on the ordinary urban and suburban landscape embodied in objects such as pylons, poles, and wires, as well as by road signs, billboards, street furniture, lighting, road markings, poorly trimmed trees, "coarse and tasteless buildings and insensitive pieces of infrastructure."[30] Thus, from Peter Blake's *God's Own Junkyard: The Planned Deterioration of America's Landscape* (1964) to Robin Boyd's *The Australian Ugliness* (1960), followed by Donald Gazzard's *Australian Outrage: The Decay of a Visual Environment* (1966), a genuine critical movement unfolded in the face of the signs of a diffuse suburban landscape, an important aspect of which was embodied in the electricity network, increasingly identified and imbued with connotation, and denouncing a poor and disorganized policy of planning the national territory.

> The Australian Ugliness is bigger and better here. . . . The ugliness . . . falls into two categories: accepted and unintentional. Australia's accepted recognized ugliness is no more than the normal blight which afflicts growing communities, especially, rich, young, industrialized, growing

John B. Jackson, sketch of a road with telegraph wires, 1947. Reproduced with permission of the Licensor through PLSClear.

communities. Part of it is the blight of age. . . . Another part is the blight of expediency: trees uprooted to save diverting a few yards of drain, the ill-considered and uncoordinated assortment of posts, hydrants, bins, transformer, benches, guards, traffic signs, tram standards, a hundred other necessary public appliances, and neons, placards, stickers, posters, slogans—all bundled together like an incompetently-rolled swag with loops and tangles of overhead wires.[31]

The identification and new aesthetic awareness of the ordinary suburban landscape is effectively a stage of the phenomenon of recovery of these objects of daily life and mark a pop sensibility that would subsequently register in postmodern discourse.[32] John B. Jackson declared in the 1950s that "U.S. Highway I is in fact one of the most sensationally ugly roads in America," adding that "its horrors end by fascinating the traveler" and comforting him during his long hours of empty desert and national prairie crossings.[33] If the founder of *Landscape* magazine—and a spokesman for the emerging concept of human geography—wondered whether we should not try to assimilate the highway landscape, and even develop "a kind of love for it," it was not until 1972 that we would witness a true heroic celebration of the media, signage, and commercial pomp of the highway with Venturi, Scott Brown, and Izenour, in their iconic book *Learning from Las Vegas*.[34] As with Janus, the Roman god with two faces, the accessories associated with the highway are simultaneously sublime and ugly, lyrical and prosaic, ordinary and heroic when one pushes the criteria of observation beyond their functional appearance. These judgments follow from one to another, superimposed from subjective visions.

Renamed "the hardware of the new world" by François Béguin, the poles, the wires, the pylons, and the antennas often remain disturbing, even in hindsight.[35] They stigmatize the landscape as "such a tangible, imperious, omnipresent and gloomy manifestation of the uniform technical order with which its destiny is henceforth associated."[36] The linear structure of the electrical networks has been seen in more recent times as a form of "planetary standardization."[37] A philosopher and geographer, Béguin became a militant critic, denouncing a bad form of progress that was materialized by the presence of technical objects in the landscape, in which the electrical network embodies the negative idea of a cutting, separating, and aggressive line: "The images that spontaneously come to mind are above all images of decomposition, the electric lines that cut through the forests, cut the line of rivers, run along the side of mountains, block the horizon."[38] From a sound that had previously been perceived as divine, for example, in Thoreau, it ends up producing a noise that screeches in the ears of the French author: "They make the perceptive field creak and grind, just as they do to the broadcasts captured by the radios on board. Forms which are always separate and distinct from others, forms without places, lines without end and without escape. . . . From now on, the pylons and the electric lines crystallize a process of protean invasion in a simple image."[39] In this highly visual evocation, the line almost becomes an organic substance, and in its excrescence, a kind of reproducible and tentacular element—a technological fractal.

The category of the ugly has traditionally linked to that of the sublime. Already Kant, in line with Burke, saw ugliness as an instrument of negative pleasure, of "the delicious terror at the origin of the sublime."[40] Here the ugly does not bear a morally negative meaning. Instead, it embodies an active principle of the sublime that, beyond the displeasure it provokes, allows reason to rise above the senses: "The whole aesthetic of the 18th and 19th centuries strives to make the ugly and the sublime two categories adapted to modern sensibility."[41] These categories make it possible to build an aesthetic theory of modernity for understanding the arts that were no longer considered "beautiful."

However, it was only in 1853 that the German philosopher Karl Rosenkranz, by pursuing Hegel, would write an essay entitled *Aesthetics of Ugliness* and try to deploy the aesthetic categories of the ugly.[42] According to the author, the beautiful can appear only if there is ugliness as well; and these two manifestations are also present in nature, in the spirit, as well as in art. He defined the aesthetic sense of ugliness according to three categories—deprivation of form, incorrectness, and disfiguration—while adding an important ethical connotation when he associated ugliness's most extreme achievements to a form of the satanic. In his essay, Rosenkranz stated: "Descent is impossible without also gaining admittance into the real hell, the hell of evil, since the ugliest ugliness is not that which in nature repels us in swamps, crippled trees, newts and toads, in gaping sea monster jaws and massive pachyderms, in rats and apes; it is the selfishness that reveals itself in spiteful and frivolous gestures, in the furrows of passion, in crooked glances and—in crime."[43]

At the height of the age of industrialization and the connection of architecture and cities to networks of convenience, such as gas, running water, and electricity, Rosenkranz addressed the "dynamic processes of nature" such as cohesion, magnetism, static or galvanic electricity, and chemical reactions, calling them "neither beautiful nor ugly" in themselves. He admitted that the effects produced by these phenomena can be beautiful, however: "the sputter of electrical sparks, the zigzag of a lightning bolt, the majestic roll of thunder, the change in colour of chemical processes, and so on."[44] It is in connection with these fantastic results that gas can generate "not only beautiful but also ugly forms," depending on the different types of deformation of its basic spherical shape. Rosenkranz does not mention the technical supports of these phenomena that he defines as natural, and he stops with the judgment of their effects.

At the end of his essay, however, he gives an ultimate definition of the beautiful and the ugly: the first associated with freedom—ethical freedom of the will, spontaneity of the intelligence, and free movement of nature—and the second defined instead by un-freedom.[45] This un-freedom entails an absence of self-determination, or a "contradiction of self-determination against the necessity of the essence of a subject." For Rosenkranz, these are the characteristics of ugliness. On the one hand, it is easy to consider the use of these terms as affording a mode of interpretation, common to architects or critical thinkers, in the face of the capitalist technical regime. One can consider the fixity—even rigidity—of modern infrastructures, along with their conditions of use, and the economic models they impose as a deprivation of freedom, falling under the aesthetic category of the ugly. On the other hand, if we take into account a certain open access to information, the capacity to communicate freely beyond physical and territorial borders, and the multiplication of exchanges that result from modern networks, we also detect elements that lead to the opposite judgment.

Whether they are considered as a subject for criticism, celebration, reverie, diversion, or poetic inspiration, energy and telecommunications infrastructures have been part of the mechanized landscape since their gradual installation in the mid-nineteenth century. Moreover, in their physical permanence and structural

Fernand Léger, *Les pylônes (étude pour La Ville)*, 1947. Musée National Fernand Léger, Biot. RMN number 15-525775. bpk / RMN—Grand Palais / Adrien Didierjean © 2023, ProLitteris, Zurich.

simplicity—often just an orthogonal grid of cables and poles that has not evolved much over time—these infrastructures still refer to the story of their initial installation. They build a bridge to the era of industrialization, with its myths and visions of progress. And it is through these material forms of technology that telecommunication weaves its most explicit links with architecture and landscape. Or rather, it becomes an infrastructure that merges completely with the landscape, instead of being detached from it, as was the case of a factory in the background of a classic landscape, for example. The landscape becomes "infrastructural in nature," because now it exists only through its vertebrae of networks and technical systems.[46]

Thus, a later representation by Fernand Léger reflects a much softer attitude toward modern infrastructures. *The Pylons* (*Les pylônes*), from 1947, are once again inserted into a rural landscape, but they no longer express the symbolic, heroic, and sublime feelings evoked by the North American images. We can see three or four cows, a gentle topography, and a slightly cloudy sky; the whole is depicted in a very graphic, childlike pictorial style, with marked contours, transparent colors, and very approximate scales, without any realistic or allegorical intention. The pylons are mesh structures that stand out on the horizon of this landscape of hills. They do not form any linear movement. Instead, they seem to zigzag in a confused way over the territory, without rhythm and without systematic logic: a grouping of three in the foreground of the image, and three others, all different from one another, in the back. One could say that this is more like an assessment or report.

Every landscape is inhabited by a pylon. Léger does not dwell on the qualities, the physical, aesthetic, allegorical, or functional attributes of these objects. It is the image of the mesh through which one perceives the landscape that interests him. It is likely that critical observation, rather than celebration, underlies the artist's intention. In an anecdote about *The Builders*, the painter observed: "There were high-voltage pylons being built near the road. Men were leaning over them, working. I was struck by the contrast between these men, the metal architecture, the clouds, and the sky. These tiny men, as if lost in a rigid, hard, hostile ensemble. That is what I wanted to render directly, without concession or compromise."[47] Between the grandeur of nature and of modern infrastructure on the one hand, and the smallness of the human being on the other, there is a contrast of scales that provokes, this time, not a feeling of sublime exaltation, but rather of discomfort and disorientation—one that will inevitably play a role in the perception of the modern mechanized world.

3 *Technics: From the Line to the Network*

An image from the American Telephone and Telegraph Company of 1917 shows the planets connected by thousands of telephone wires. It brings to mind one possible antecedent—an engraving by the French cartoonist Jean-Ignace-Isidore Gérard, better known as Grandville, from his lithographic collection entitled *Un autre monde*.[1] In both illustrations we find the idea of a modernized cosmos connected by bridges and wires—visions communicating a concept of the world unified by a network that transcends all scales, to the point of encompassing and connecting all the planets together. The vision of Grandville, French draftsman and ancestor of Surrealism, remains based on the idea of a strong, monumental, ornamental, and rigid infrastructure, while the advertising image for AT&T emphasizes an idea of fluidity and the technical simplicity of a modern communication network that opens up new horizons.

With the invention of the telegraph in 1837, the telephone in 1876, and the radio in 1896, urban and rural landscapes underwent a radical and permanent transformation. The technological elements that make up telecommunications systems—the poles, wires, antennas, transmission towers, and the like—became universal objects and signs, common to the whole planet. These physical landmarks played a role in the transformation of everyday life: first, by creating instant communication networks on a global

AT&T advertising, 1917. N. W. Ayer Advertising Agency Records, Archives Center, National Museum of American History, Smithsonian Institution.

scale, contracting the perception of distance and speed, and second, as equipment around which cities were to be redesigned, thus generating new models of spatial organization.

While cities went through a process of densification during the industrial revolution, the modern period of the 1920s and

Illustration from Grandville, *Un autre monde* (Paris: H. Fournier, 1844). Getty Research Institute, Los Angeles (92-B7023).

1930s was marked in the United States by a process of dispersion: the phenomenon of urban sprawl. The transport and telecommunications infrastructures allowed a kind of opening up of space—thanks to the construction of roads facilitating the rapid movement of individuals, as well as the penetration of otherwise inaccessible territories. They also contributed to forging the modern myth of the contraction of space-time through the phenomenon of instantaneous communication. Alexander Graham Bell, the inventor of the telephone, was "the man who contracted space," according to his biographer Catherine Mackenzie.[2]

This idea was also the focus of marketing campaigns by the American Telephone and Telegraph company, which emphasized properties of this powerful network as a "space annihilator" that could magically reduce the immense American territory to the scale of a neighborhood.[3] These various fantasies of spatial collapse through the introduction of modern technologies were integral parts of publicity for the telecommunications and transportation companies, and they were crucial for media theory as well.[4] Naturally, this expanded usage of the notion of space entailed a loss of its real constitutive features. Instead, space frequently became the support for a kind of allegorical vision that linked the introduction of these modern technologies and the distribution of the different objects that compose them.

AT&T advertising, 1909. Courtesy of AT&T Archives and History Center.

AT&T advertising, 1930. Courtesy of AT&T Archives and History Center.

AT&T advertising, 1910. Courtesy of AT&T Archives and History Center.

Wolfgang Schivelbusch, for example, describes how the railroad provided easy access to previously impossible spaces, but he also speaks of a metaphorical "destruction" of its essence through the reduction of the experience of travel to points of departure and arrival.[5]

This chapter explores how the development of telecommunications infrastructure resonated in modern urban theory, specifically in the discourse of American territorial planning in the 1920s and 1930s. The aim here is to analyze how professional engineering practices were absorbed by modern architects and planners and synthesized into a new vision of the city as well as the countryside. Two significant figures will be studied in particular: the historian Lewis Mumford and the architect Frank Lloyd Wright. Both Mumford and Wright considered assimilation of the modern mechanized territory as pivotal for future urban development. In both critical discourse and their projects, Mumford and Wright used telecommunications technologies, among others, to turn against the condition of modern cities. Equally opposed to the unplanned growth of the metropolis, they proposed alternative visions to the emerging phenomenon of sprawl.

Regionalism was the guiding principle in the work of Mumford and his colleagues at the Regional Planning Association of America (RPAA) during the 1920s. Mumford was strongly influenced by the pioneering urban planning theories of the Scottish biologist Sir Patrick Geddes, who observed the territory by studying the complex interrelationships between man and his environment and introduced the term "region." Geddes went beyond the opposition between town and country and looked at the territory in terms of continuity, with the town being a more concentrated form of settlement resulting from activities initially practiced in the countryside. Geddes proposed a new development model, the city-region, which encompassed the social, economic, and material life of a place.[6] First explained in a 1905 publication, the Valley Section method illustrated how "natural occupations" such as hunting, mining, or fishing were supported by physical features of the land that, in turn, determined patterns of human settlement.[7] This use of the section in particular, enabled a synchronous vision of the different resources and activities co-present on the territory and an understanding of different forms of organization. It was from the same understanding of a region as a geographic area unified in climate, soil, vegetation, industry, and culture, and borrowed from the discipline of geography, that the RPAA built its precepts of regionalism. "The regionalist attempts to plan such an area so that all its sites and resources, from forest to city, from highland to water level, may be soundly developed, and so that the population will be distributed so as to utilize, rather than to nullify or destroy its natural advantages."[8] This conception of territorial planning anticipated ecological thinking and bioregionalism. It also featured a form of confidence and willingness to assimilate technological progress.

As a historian and philosopher of technology, Mumford believed that technology could lead humanity to a better future. In his progressive view of history, regionalism was part of a natural process and represented the direction of future urban development.[9] His reading of American history was constructed through several periods of occupation and migration: from the movement of settlers across the land to the concentration of goods and capital in the financial centers of the country. According to Mumford, these had created unbalanced and unhealthy forms of land appropriation that could only be resolved by the "technological revolution" that was at the root of what he called the "fourth migration." In his discourse, Mumford remained very attached to the physical role of infrastructures and their technical essence, as primordial elements in the radical transformation of society. Thus, in the field of transportation, he observed the way in which the

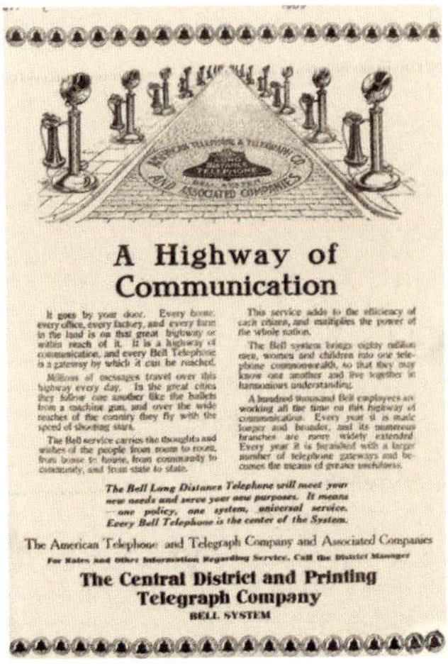

AT&T advertising, 1909. Courtesy of AT&T Archives and History Center.

AT&T advertising, 1919. Courtesy of AT&T Archives and History Center.

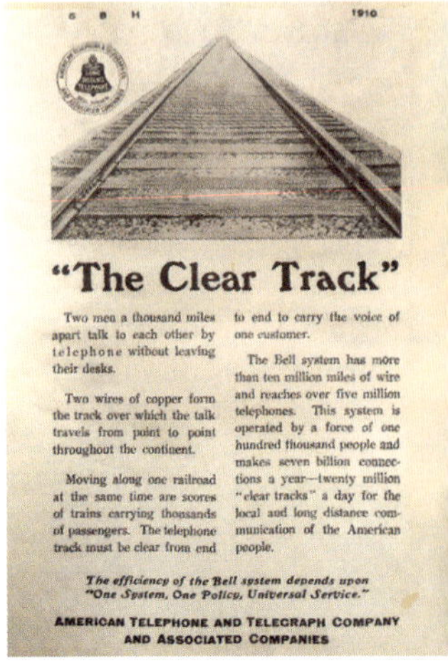

AT&T advertising, 1911. Courtesy of AT&T Archives and History Center.

automobile offered a new capacity for movement that would reinscribe the territory. The strict linearity of the railroad and the train was supplanted by the more autonomous and free trajectories of the automobile, capable of reaching larger parts of the territory, with a more diffuse movement of goods and markets. This technical decentralization thus enabled a substantial deconcentration of urban populations. In an even more incisive way, the installation of telephone lines represented for Mumford a social revolution, opening up communication between people located at a great distance from each other. The urban concentration of industrial societies was the condition before the invention of the telephone, when every direct exchange had to take place on the spot. The highway system and the telephone offered better means of communication than their predecessor technologies, the railroad and the telegraph, which still produced concentration effects: "The telegraph symbolically follows the railroad; the telephone, with kindred symbolism, follows the motor highway. So much for the business end of communication."[10]

These couplings of technical infrastructures became emblematic elements of the modern landscape that appeared very frequently in the American pastoral representations of the time, as well as in the publicity images of the American Telephone and Telegraph Company, which went so far as to propose several

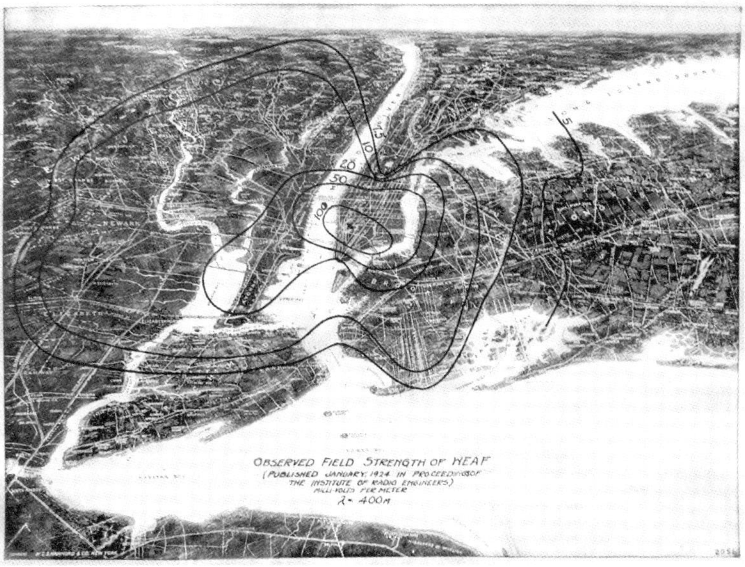

Field of intensity of radio transmission over Manhattan and surroundings. Published in January 1924 in the *Proceedings of The Institute of Radio Engineers*. George H. Clark Radioana Collection, Archives Center, National Museum of American History, Smithsonian Institution.

metaphors and analogies. With figurative expressions such as the "Communication Highway," the "Bell Highway," and the "Double Road," AT&T highlighted the scope of the telecommunications infrastructures and identified them with the more physically imposing roadway networks. Thus, objects like the highway and the railroad that evoke the movement of goods and people would be symbolically identified with the invisible transport of messages through electrical wires. In the modern era, this complexification of exchanges materializes in different forms and combines superimposed infrastructural systems that are not independent of each other in reality—and continues to this day with the assimilation of the digital network.[11] As systems, telecommunications networks necessarily involve a movement of individuals and machines; consequently, they do not subsist independently from transportation.

Going back to Lewis Mumford, "the radio is potentially a distributive and decentralizing agency."[12] The music, news, and gossip once heard in a public square were now available to anyone who owned a radio. Electromagnetic waves radiate out from transmission points, towers and antennas, spreading over geographical areas—specifically, the regions.

Combinations of transportation and communication technologies had reduced the advantages of metropolitan concentration to zero. The physical distance between people was no longer a measure of the time required for them to communicate. Sending a letter from a house in the country became much faster when the letter carrier came right to the door with his private mode of transport, while the city dweller had to go to the post office. The task of the fourth migration was therefore to gather and channel the new possibilities afforded by modern society, and to reconstitute a healthy and happy environment for the inhabitants. As a response to these initial reflections, the regional urban organization envisaged a constellation of connected cities, separated by parks and agricultural areas. Each city was to have an autonomous life, with its own local institutions, while also specializing in some aspect of social and cultural life (museum, university, radio station).[13]

In fact, the regional model has its roots in Ebenezer Howard's theory proposed in *To-Morrow: A Peaceful Path to Real Reform* of 1898.[14] Howard's model combined the attraction or "magnet" of the city with the benefits of the countryside, through the formation of small urban nuclei scattered throughout the countryside. In the introduction to the 1946 edition of *Garden Cities of To-Morrow*, Mumford compared the invention of the airplane to Howard's urban contribution: "The former gave man wings, the latter gave him a better place to live when he came down to earth."[15] For Mumford, the regionalist model was a new technique for the age of mechanization.

The development of regionalist urban theories occurred in parallel with the evolution of electrical transmission systems. The phase that the technological historian Thomas Hughes refers to with regard to regional generation and distribution corresponds to the third stage in the process of electrification, coming after the "Edison direct current" and the "universal" systems.[16] While the first of these systems was characterized by homogeneity of supply and load and existed for only one function (a standard voltage transmitted and delivered to the customer in the same standard voltage), the universal system was characterized by greater heterogeneity. Thus, the latter brought together various types of generators (alternating and direct current), combined them in a single transmission system, using couplers (transformers, synchronous and motor generators, frequency converters), and connecting different types of loads (lighting and electricity, stationary or traction power) as well. The "culture" of this technological system was guided, according to the author, by the determination on the part of engineers and entrepreneurs to assimilate technology into the economy, with the goal of generating a universal and expanding system. This system achieved its full momentum in 1890, when the invention of the steam turbine allowed for more efficient and profitable rhythms and modes of production and led to very dense concentrations of working-class populations, at least at first. The "culture of the regional system" followed the extension of high-voltage lines during the 1920s and 1930s as well, enabling the spread of new models of urban sprawl and the expansion of the American suburb: "Point-to-point transmissions grew into networks, grids, or rings as the long lines were punctuated by major switching stations where other generating stations fed power into the system or power for nearby loads was taken out."[17]

The regional system would thus be characterized by an even greater heterogeneity of supplies and loads, articulated in a more autonomous way than the previous system, in a logic of mixed economy that stimulated the inclusion of local energy sources (different types of coal and natural hydraulic movements).[18] The line, the infrastructure of the electrical system, evolved enormously during this period of regional expansion, with the installation of steel overland transmission towers, copper or aluminum lines supplied with expensive insulation, and a complex array of lightning protection devices, circuit breakers, switches, transformers, and synchronous generators. The levels of national investment in the United States during the 1920s exceeded the expenditure on the railroad system during its period of maximum expansion.[19] The telephone network followed the same path, even though its system supported a single function and was based on more homogeneous technical constituents. The transmission of sound by electricity requires alternating current, according to the very nature of the sound vibration, which leads more easily to variation and loss of clarity: "As it is our aim to make the sound at the receiving end as nearly as possible an exact reproduction of the sound at the transmitting end, it is necessary that the apparatus for transforming sound-waves into electric waves, and vice versa, should be such as to cause the least possible change in form and amplitude; and in addition to this, it is necessary that the line should be so constructed that the current waves may be propagated with the minimum alteration."[20]

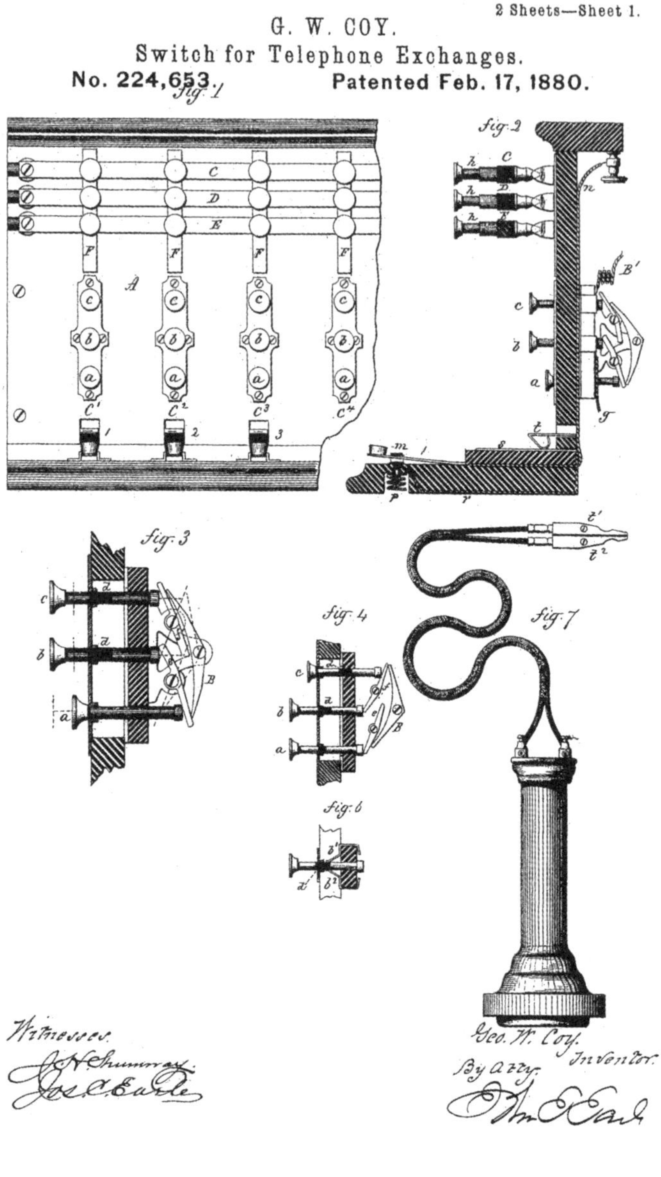

George W. Coy, Switch for Telephone Exchange, no. 224,653 patented February 17, 1880.

Simply explained in Hopkins's 1899 manual, the objective of alternating current telephone transmission was to recreate the sound emitted at the beginning of the transmission journey through wires and resistors, with the same rhythm, same tone,

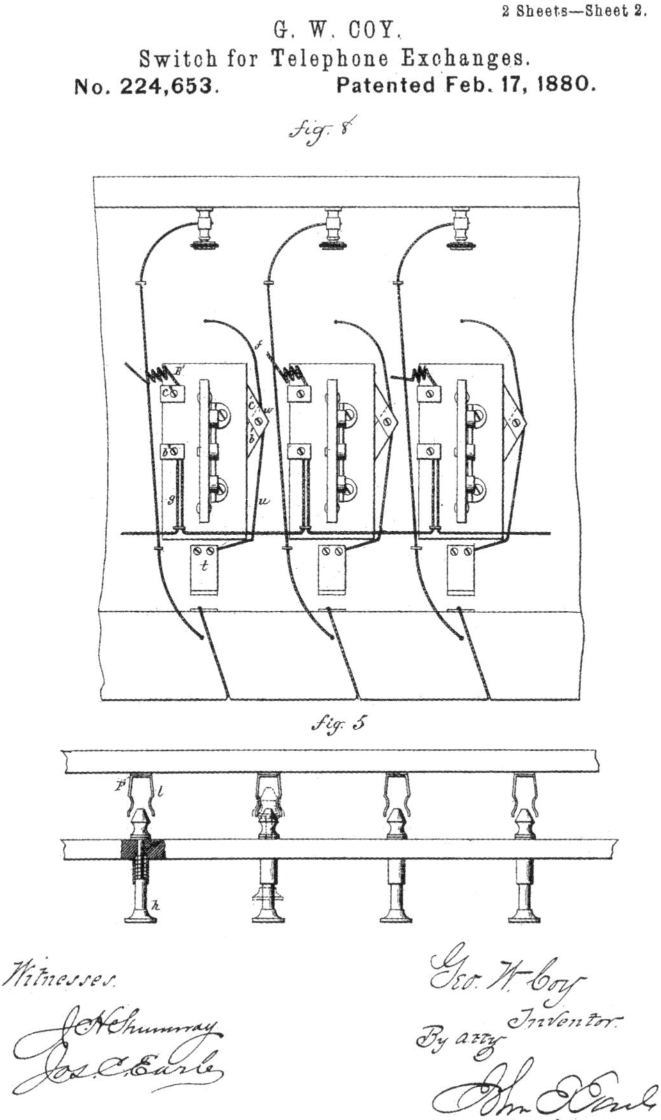

same timbre. An "impossible" goal, as the author says, but one that nevertheless guided the search to achieve the minimum possible alteration. The first permanent outdoor line was built in 1877 to connect an electrical manufacturer's store at 109 Court Street in Boston to its owner's home in Somerville, about three miles away.[21] But it was not until the installation of the first commercial telephone exchange, in a store at the Boardman Building in New Haven in 1878, by the engineer, mechanic, and contractor George Coy, that telephone communication, previously possible only between peers, became an open and expandable network. It was this possibility for articulating the system that would lead to the unprecedented success of the telephone, which unlike the telegraph that preceded it, would allow the direct transmission of voice sound.[22] After listening to a lecture by Bell in New Haven in 1877 that included telephone calls to Hartford and Middletown (two other towns in the state of Connecticut) as a demonstration, Coy bought the telephone license to establish his own company, the District Telephone Company of New Haven, building on the possibilities created by his invention of the standard.[23]

The telephone exchange, like the telegraph, was based on a network that allowed the interconnection of more than two remote users. The manual switchboard was at the heart of the telephone exchange. It served as the hub for all telecommunications. Using electrical cords or switches, the switchboard allowed the connection to different lines through a manual procedure, performed in the central offices. In other words, thanks to the switchboard, the number of lines (each pair of connected telephones constituting a line) could be reduced thanks to the articulation of a branched network.

After an initial period of competition between several local companies, the Bell Telephone Company became the American Bell Telephone Company in 1880, and it subsequently pursued the goal of a national network with the creation of the American Telephone and Telegraph Company in 1885, explicitly dedicated to the project of installing a commercially viable infrastructure on a national scale. The first long-distance service, between New York and Chicago, was established in 1892, and the first transcontinental service, between New York and San Francisco, in 1915. These large linear connections, combined with shorter radial extensions locally, ended up forming a complex sprawling network from different intersections and centralities. Starting in 1907, under the initiative of director Theodore Vail, AT&T invested heavily in perfecting the long-distance network—by incorporating local companies with the goal of constituting a "universal system": a national service at a moderate price. AT&T then attained the status of a natural monopoly, which it maintained until 1982.[24]

But countries that engaged in this race to modernize were not uniformly equipped. In fact, there was a big difference between the condition of a city user and a country user. Compared to other European countries, or even to New Zealand, for example, the American rural dweller lived in a backward situation.[25] Before the creation of the great governmental programs, such as the Tennessee Valley Authority (1932) and the Rural Electrification Administration (1935), the American countryside was virtually excluded from the country's extraordinary modernization process. Private companies did not naturally invest in rural areas because they were less profitable, due to the lower concentration of inhabitants and the need for larger-scale infrastructure scattered over the territory. For private entrepreneurs, communication was initially seen more as a commodity than as a social service.[26] The situation would change to a great extent during the 1930s, when electrification became a social program in the United States with the development of large-scale infrastructure on a national scale.

One of Roosevelt's programs, the Farm Security Administration (1935) included a photographic report on the American countryside, commissioned from several photographers including Russell Lee, Walker Evans, and others—and making use of their sensitive views of the dual effects of modernization. Some of these images highlight the implementation of electrical infrastructures, but a sense of desolation and emptiness always outweighs these technical systems. What is striking in Russell Lee's image is the repetitiveness of the houses in the street of a rural village, punctuated by electric poles. In Evans's framings, the orthogonal intersections of electrical infrastructure become the grid, the observer's point of view of inanimate urban or semi-urban situations that contrast sharply with a sublime vision of the technical landscape.

Mumford and his colleagues had actually anticipated this renewed interest in rural America, and they directly inspired the programs promoted by Roosevelt's administration, particularly as several members of the Regional Planning Association of America were personally invested in the New Deal programs. Regionalist theory should not be confused with a return to a purely pastoral ideal. Instead, it represented the emergence of a new form of decentralized urbanization, uncompromisingly based on a

Photograph by Russell Lee, n.d. Canadian Centre for Architecture, PH1985:0302.

mesh of modern infrastructure. The members of the RPAA were opposed to the planning of metropolises, which combined with capitalist interests to generate the deplorable condition of "metropolitanism."[27] Instead, they advocated and affirmed a new culture that would begin with the migration from the metropolis to more balanced centers and communities between rural and urban conditions.

In 1923, New York Governor Alfred E. Smith asked the RPAA architect Clarence Stein to head a planning commission for New York State. The final report submitted in 1926, the *Report of the New State Commission of Housing and Regional Planning to Governor Alfred E. Smith*, illustrates the application of regionalist theories, with significant public intervention, for the development of New York State.[28] This study should not be seen as a plan, but as an analytical collection of data, for use in future planning of the territory.[29] Beyond local resources, the study addressed the ongoing technical evolution and transformation of the region through new infrastructure: the replacement of steam power by electricity, the development of public and private transport, and the expansion of the road system. This efficient and economical use of technology led to new forms of decentralization. The study

Photograph by Walker Evans, Reedsville, West Virginia, 1936. Library of Congress LC-DIG-fsa-8c52037. © Walker Evans Archive, The Metropolitan Museum of Art.

was largely ignored at the time of its publication, but it remains a clear example of the application of the regionalist model.[30] The diachronic succession of maps, the result of several layers of analysis of the territory, graphically shows forms of expansion and contraction of networks, resulting from the evolution of techno-economic modes of production as a function of local resources. A first phase of homogenous sprawl that corresponded to the beginning of territorial exploitation (1840–1880) was followed by a second phase of concentration along the central railroad routes (1880–1920). With the transformation of factories by the steam

Photograph by Walker Evans, View of Morgantown, West Virginia, 1935. Canadian Centre for Architecture, PH1984:0499. © Walker Evans Archive, The Metropolitan Museum of Art.

engine and the agricultural competition with the American West in combination, this phase concentrated growth along the Central Valley belt of the state. The power of modern technology, the development of the road network, and the impetus toward a national vision of the territory resulted in a detachment from the local constitution of the territory, which would become explicit in the diagram by a linear concentration, directed toward more distant horizons, at the expense of small local centralities. The third phase advocated by RPAA members, was to develop along a powerful, complex, and hierarchical system of networks. The map drawn shows various urban nodes that tap different local sources across a territory marked by a multiplicity of points. This is the second industrial era, marked by the advent of the automobile, a strong road system, and electrical transmission lines that allowed the effects of modern society to be channeled through a new form of encounter with the local sources of the territory. Graphically, the map shows a variety of features, lines, meshes, and points that restore a thickness to this representation of a territory that had been disembodied from its physical constitution

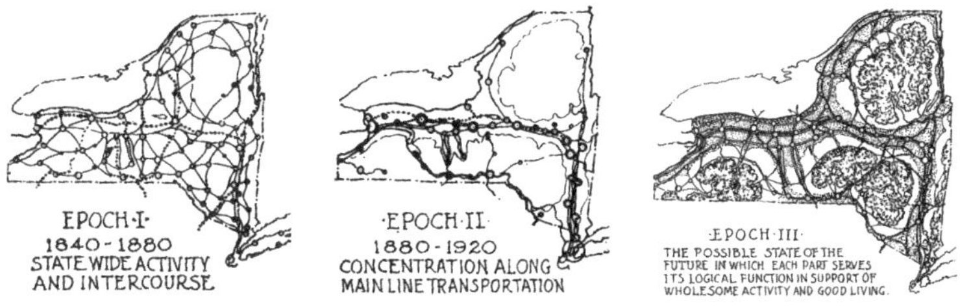

RPAA, Regional Plan for the State of New York, 1926. Published in *Report of the New York State Commission of Housing and Regional Planning to Governor Alfred E. Smith* (Albany: J. B. Lyon and Co., May 7, 1926).

by an uncontrolled phase of the modernization process applied to the territory.

This rejection of metropolitan density was also shared by Frank Lloyd Wright. In his quest to identify what American modernity expressed, this ingenious architect believed that the overdeveloped modern city reflected a dangerous interference between political and economic power.[31] Wright was particularly opposed to the concentration of these powers in the hands of what he called "capitalist fools." His opposition to the exploitative capacities of industrial production and capitalism was the basis for his Broadacre City project, a plan for a new type of urban community that he developed between 1932 and 1958. Through the combination of two principles—"Democracy" and "organic" architecture—Wright imagined a different and idealized version of the American suburb. Broadacre City embodied the concept of a continuous system of urbanization based on a one-acre-unit module. The project followed in the footsteps of the Jeffersonian tradition, which considered agriculture as the primary economic force that would enable the American individual, the landowner and farmer, to become the model city dweller. Wright would subsequently organize this agrarian space by reintroducing many urban functions.[32] For Jefferson, America was "the world's largest city," and if America was also "the garden of the world," Wright would attempt to give it "physical form" through planned distribution.[33] The Broadacre City project thus combined several founding themes of nineteenth-century American culture—while also relying on progress and the means offered by modern technologies. The cultural frame of the project was also nourished by the thought of transcendentalist poets such as Ralph Waldo Emerson and Henry Thoreau, who saw communion with Nature as a form of reconciliation of the individual with the divine, and also as a defense of Democracy. In the Jeffersonian grid, with its American literary proponents, such as Walt Whitman, Wright found a symbolic and concrete figure for an equitable distribution of goods and opportunities—geometric spatial translation that accommodates a great concentration of values and techno-social connotations. The factory, the farm, the office, the store, the church, and the theater were all located within ten miles of the Usonian, the individual house of Broadacre.[34] Not only was this urban organization physical, but it also involved a redistribution of goods and capital, thus eliminating the social differences between the country and the city. Moreover, unlike Jefferson, who through a cadastral and rural reading of the American land wished to anchor the principle of democracy in the country, Broadacre's grid also physically integrated the linearity of modern infrastructures.[35] If on the one hand, the new standard measure for Wright was "the man sitting in his automobile," on the other hand, the communication and information networks—"radio, telephone and telegraph: electrical intercommunication becoming complete"[36]—were postulates for urban, social, and economic development toward an idealized "Free City," the "City of Democracy." Transport and communication were still at the heart of an urban planning model that aimed at a social ideal of well-being and equity.

Much has already been said about the construction of the American territory by systems of lines. This geometric framework is also explicitly present in the architect's descriptions of an evolving horizontal grid, emblematic for him of a new society: "Horizontality has received an impulse that widens human activities immeasurably."[37] This horizontality, which again embodies the idea of conquest and territorial expansion, is a figure that often slips from a symbolic reading to a physical materialization, as in drawn views of Broadacre City, which propose yet another type of horizontality, stemming from the sprawl of the city-countryside at ground level, without otherwise allowing the presence of linear aerial infrastructures to appear, but promising a physical absorption of their technical capacities. It is interesting to note that the documents Wright offered to illustrate Broadacre were always zenithal aerial views, or even more often bird's-eye views—photographs of the model and several perspective drawings of the city—that offered, with the many gridded planes, a top-down and oblique perception of the project. The representation of a horizon that is never closed but always expanding, the presence of the sky and its flying objects contribute, as with the figure of the grid, to the idea of continuous planning, without borders, that the architect particularly cherished.

Now, if the horizon separates transparency from opacity, matter-earth from space-light, it becomes truly linear for Paul Virilio with the introduction of the Albertian perspective.[38] This technical transformation of representation implies an important philosophical change, because compared to the previous Sienese techniques that made the horizon adhere to the things represented, the drawn space precedes the objects that occupy it. The space becomes an entity that includes the spectator in a scene that could have taken place before him.

Similarly, the grid, according to Bernhard Siegert, is a cultural technique that lies between mathematics, topography, geography, and politics. It is a diagrammatic procedure that allows for the representation of a three-dimensional geometric space on a two-dimensional plane in which objects can be placed (and subjective views, potentially without spatial limits, can be constructed). This technique allows one "to store data that can be implemented in the real as well as in the symbolic."[39] For Rosalind Krauss, this set of intersections of lines on a plane imposes itself in art as a figure of modernity by its anti-mimetic, anti-natural spatial condition, and by its relation to the present time, which makes an abstraction of past traditions.[40] Thus, for modern cubist artists and for De Stijl, Mondrian, Malevich, and others, the grid is a contestation of the dogma of imitation in art. It denounces an absence of narration: "they landed in the present, and everything else was declared to be the past."[41] This conquest of autonomy in art, which is moreover widely debated,[42] nonetheless reflects a possible interpretation of the values embodied by the straightness of the sign. Geometric and moral rectitude—it is this simple rigor founded on a conception of progress and technical evolution proper to the modern era from which Broadacre City's plan could, among other things, draw inspiration. Thus, when we return to the materiality of infrastructures and architecture, we notice that the transition from sign to concept, and even to object, becomes a constant exercise that requires several possibilities of reading.

In spite of a common desire on the part of both Mumford and Wright to propose models of decentralization of the American territory based on principles of social and economic redistribution of goods, and relying on the progress of modern distribution technologies, one can also note important differences. First of all, these oppositions are also due to the different training of the two thinkers—one a theoretician and the other an architect—that are formally expressed in the adoption of a branched, irregular system, following the geography of the regional network territory on the one hand, and the more abstract and geometric diagram of Broadacre City's generating principle on the other. But it is in Mumford's rich correspondence and critical activity that we can reconstruct the relationship of friendship and great respect between the two men, who were also able to assimilate direct criticism of each other's visions. "Broadacre City, as Wright has conceived it, is both a generous dream and a rational plan,"[43] said Mumford in 1935, in a New Yorker article entitled "Sky Line." Further on, he states that "the weakest point in Broadacre City is the design of the minimal house. Wright, who hates for the lower-income groups which compares very unfavorably, I think, with the European and American 'housing' he detests. He should have permitted himself to dream more generously. As soon as he reaches the level of the 'two-car' house, he shows that he is still our most original architect, as capable of producing beauty in standardized units as in his favored materials that spring out of the earth."[44] A lively epistolary exchange would follow, with Wright refusing to see any value in Mumford's comparison with "the German tenement and slum solution," and Mumford defending the possibility of changing forms for communities of medium and variable density, and therefore opposing the idea of Broadacre as a "universal solution."

Finally, although Mumford does not express it directly about Broadacre City, in a passage from *The Story of Utopias* of 1922, we find a critique of the orthogonal system of the city of Coketown, the fictional industrial city described in 1854 by Charles Dickens in the novel *Hard Times*. The radial system, taken from the system of "intercommunications," is for Mumford much more economical than drawing straight lines forming rectangles: "it is impossible to approach from the front the jails, hospitals, and sanatoria of which Coke-town boasts; the tendency is to run past them. So much for the physical layout of the industrial city; what remains is obscured by smoke."[45] The grid system is for Mumford clearly associated with the image of the industrial city, while the branched network of telecommunications represents the phase of regionalist emancipation that he defends.

AT&T advertising, 1911. Courtesy of AT&T Archives and History Center.

As observed here by the examples of Lewis Mumford and Frank Lloyd Wright, the territorial planning theories in the 1920s and 1930s attempted to situate the emergence of the technologies of communication technologies in a political and social perspective. Both Mumford and Wright believed in an idea of progress that was different from the one that generally accompanied the new "machine age." If modern technologies were a means to achieve new urban centers, they also constituted a kind of model for their constitution and their functioning. The way in which telecommunications infrastructures developed in the United States during the period of the regional system, making networks more complex and multiplying centralities, contributes to the generation of new city models. The electrical system, the telephone and the radio all developed considerably during the 1920s and 1930s, based on systems of interconnected lines and cores (the power station, the switchboard, the station, the antenna) that could be used both for long-distance communication and to provide service to a region. The progressive ideal of equitable distribution and free access to communication services—factors often associated with the American principle of democracy—would then be largely associated with the dissemination of infrastructures on the national territory.

Telecommunications networks are often equated with the notion of "flow." By traveling through the ether, they convey a sense of lightness, of invisibility, especially when compared to the "heavy" materiality of transport infrastructures. Yet their physical presence and organization have not been without impact on the constitution and organization of modern cities and, even more significantly, on the structure of societies.

The case of Los Angeles is useful to illustrate, through the prism of the urban theories mentioned, the relationship between the regional-scale evolution of a city and its large agglomeration, and the development of the telephone infrastructure. According to historian Emily Bills, the telephone was the primary infrastructure that united the Greater Los Angeles area into a single, multinucleated urban entity.[46] The first telephone company was established in 1881, but by 1895 Los Angeles had become the metropolis with the largest number of telephones per capita.[47] Although its development had been preceded by the telegraph and the railroad, none of these services had developed as rapidly and as widely. Also, the telephone opened up the possibility of establishing connections between the city's downtown, where the administrative headquarters were concentrated, and the places of economic production scattered throughout the region. Moreover, in the early stages of its evolution, the telephone infrastructure facilitated the development of cooperatives and regional economic nuclei through the formation of local telephone companies that allowed them to develop their businesses independently and competitively from the economic center of Los Angeles.[48]

For the historian, the example of Greater Los Angeles allows us to observe how several economic and technical issues had a significant impact on the physical development and planning of the city and its region. The evolution of the telephone infrastructure has contributed decisively to the urban constitution of this territory and to its modernization process and, conversely, this particular environment made of economic, social, and technical conditions has allowed a rapid and assertive development of this infrastructure.

The system of functioning, distribution, and implantation of modern infrastructures does not follow in a univocal way a single economic, technical, political, or social project, but it is true that one can recognize certain preliminary objectives in the form of a network. Nowadays, the term "topology" is often used, introduced in the middle of the nineteenth century to designate a mathematical approach to the "study of place."[49] Applied to infrastructures, topology allows us to find the appropriate layout of networks, their points of distribution, and their modulation of activities—their standards. Topology comes to define the organization of a system: the combination of connection and distribution nodes. Depending on the technology and the business model adopted, networks can be linear, multi-core, radial, serial, or parallel. "Cloud," "hub," "web," "spoke" are all terms that by analogy describe the shape of a network that can, among other things, refer to the way in which the authority manipulates, circulates, or concentrates information. The topology alone does not determine the layout of a system; it composes with all the components that generate the installation of a network—this same network that, in AT&T's fantastic vision, ended up encircling the entire planet.

4 *Imaginary: Eutopia of Connectivity*

"I believe that in the future wires will unite the head offices of Telephone Companies in different cities and a man in one part of the country may communicate by word of mouth with another in a distant place."[1] On March 25, 1878, only two years after the invention of the telephone, Alexander Graham Bell imagined an imminent future when communication between people located in the four corners of the Earth would be possible, thanks to a dense network of telephone wires.

The telephone was still in its infancy, capable of connecting only at the local level, but Bell already imagined that his invention—as a device, a machine, and a network—would become an intangible bridge between cities and countries, even between people and nations. Comparing the telephone to the telegraph system that was much more advanced at the time, Bell insisted on the universality of his instrument due to the immediacy of the speaking voice, which obviated the need to translate electrical signals into meaningful words. The telephone's ease of use and the increase in intimacy it promoted between people consolidated Bell's idea of a possible universal and universalizing diffusion of this technology, based on the installation of a robust infrastructure spread over the Earth's surface. Trying to sell his invention to an English company, Bell proposed developing a centralized system connecting different places to central stations, capable of distributing each communication. He predicted that as gas and water networks had so far designed the basements of contemporary cities, one day telephone wires "could be laid underground or suspended overhead, communicating by branch wire with private dwellings, Counting Houses, shops, Manufactories, etc."[2] Further on in the same letter, he remarked: "I am aware that such ideas may appear to you Eutopian and out of place, for we are met together for the purpose of discussing not the future of the telephone but its present."[3] The English company never responded to this letter, so Bell himself eventually took the initiative of creating his own company, the one that would develop the American telephone system.

Bell's ideas were obviously premonitory of the beginning of the "age of telecommunication," while expressing a crucial distinction: he was talking about a future that he defined as Eu-topical, not U-topical. Bell believed that the ability to connect across long distances would create an Eutopia, a true region of possible well-being and order, and not the idealized and abstract "non-place" expressed by the notion of Utopia. This distinction reveals Bell's belief in the power of telecommunications as a device for social cohesion. He was convinced that a better future could be achieved by reforming society from within, through more direct means of communication, not by reinventing a new world. Bell concluded his letter thus: "By bearing in mind the great objects to be ultimately achieved, I believe that the Telegraph Company can not only secure for itself a business of the most remunerative kind but also benefit the public in a way that has never previously been attempted."[4]

Bell's techno-optimism should not be seen as the position of an isolated inventor. It was part of an emerging ideology shared by other key figures of modernity. Parallel to the development of the telephone, the western world was invested in a vast process of electrification that caused a radical transformation of human horizons, both real and imaginary. The diffuse, planetary, and permanent installation of cables, wires, and poles was going to make a machine in the garden—to take up Leo Marx's efficient image—announcing a modern and mechanized society.[5] Despite its constant and universal presence in our daily lives, this landscape of networks, which has become so ordinary to us, represents the dream of a conquest of natural territory by technology, the transmission of sound, energy, and information across a potentially infinite space. These ideas have produced several utopian

Film still from *Le Mélomane*, by Georges Méliès, 1903.

scenarios that heralded the advent of the twentieth century. This techno-utopian tradition fluctuates easily between reality and imagination, developing an extremely fertile creative terrain.

Considered one of the first directors of the science fiction genre, with his 1902 film *A Trip to the Moon* (*Le voyage dans la lune*) Georges Méliès leaves us with a suggestive testimony of how the increased capacity of signal propagation of modern telecommunications has inspired several aesthetic rewritings of the modern landscape. In his 1903 film *The Melomaniac* (*Le Mélomane*), Méliès evoked the imagery of modern telecommunications through the animation of the wires, cables, and poles that became the central figures of his story.

The foreground conveys the magical dimension of the network and the musical fantasies that it inspired, in a kind of montage and detour with a Dadaist foretaste.[6] The sublime dramaturgy that characterizes other aesthetic evocations gives way here to humor and creative imagination. The electrical network becomes a musical staff on which an extravagant conductor inscribes his own music by adding the treble clef, the different notes, and their rhythmic values. The background shows a single house isolated in the countryside, the infrastructure resonating with this very simplified rural environment.

In his introduction to *Urban Utopias in the Twentieth Century*, historian Robert Fishman identified three urban architects who worked on an ideal, technologically advanced city, combining the power and beauty of modern technology with ideas of social and economic justice: Ebenezer Howard, Frank Lloyd Wright, and Le Corbusier.[7] Their fascination with technology was profound and selective. Although they still observed it through a nineteenth-century filter, they believed that modern technologies had overcome the old social order, albeit initially by producing social chaos, injustice, and conflict. In their ideal cities, however, technology would have repaired this degradation of modern life. They believed that an industrial society could become fundamentally harmonious through forms of organization that would eradicate conflict and bring order, freedom, prosperity, and beauty. The interest in the utopian tradition was first made explicit in the case of Howard, who had been “electrified” by reading Edward Bellamy’s utopian novel *Looking Backward*, published in 1888 in the United States. Today it is still considered a visionary precursor to many modern inventions, such as the credit card, electricity for heat and light, and the “musical telephone.” In a 1910 article entitled “Spiritual Influences Towards Social Progress,” Howard spoke of how Bellamy’s utopia was an inspiration for thinking about a nation capable of securing “order, justice, unity and benevolence.”[8] He personally lobbied American publisher William Reeves to produce the book in the United Kingdom, committing himself to the acquisition of the first hundred copies and working to write the index to the book free of charge. Despite the centralized and totalizing nature of the “urban” society described in the novel, Howard was fascinated by the idea of a just world based on a socialist model and became involved in the Nationalization of Labour Society, founded in 1890 to promote Bellamy’s ideas.[9]

The objective of this chapter is therefore to study the way in which telecommunications technologies such as the telegraph, the telephone, and the radio entered into the discourse of modern thought by forging the organization of cities of the future, whether utopian or eutopian. From the study of three main figures, we observe different ways of assimilating the technological landscape in their projects. These are Lewis Mumford (who in

Edward Hopper, *Road in Maine*, 1914. Whitney Museum of American Art, New York, USA. © 2025. Digital image, Whitney Museum of American Art / Licensed by Scala. © Heirs of Josephine N. Hopper / 2023, ProLitteris, Zurich

an even more explicit way than the designer of the garden cities returns to the role of remote communication), Frank Lloyd Wright, and Le Corbusier. Their various perspectives also reflect a tension that appears between the American and European avant-garde of the time. Mumford and Wright applied, as we have seen, principles of urban sprawl onto a territorial scale and valued individual initiative, while Le Corbusier relied on a concentrated model of technocracy.

In a figurative comparison, two almost contemporary works from the beginning of the twentieth century illustrate opposite visions of the electric wire landscape: first, Edward Hopper's American Realism of *Road in Maine* (1914), which shows a deserted road accompanied by a line of rudimentary wooden poles, standing amid a rugged, natural landscape that turns toward a distant horizon, which is concealed by a rock.

Umberto Boccioni, *Elasticità*, 1912. Museo del Novecento, Milan.

And second, with *Elasticity* (*Elasticità*) painted by Umberto Boccioni in 1912, we encounter a much more chaotic, urban representation that confuses the different planes of the scene in a kind of dynamic vorticism. In the foreground, we can make out a knight in motion against the background of an industrial landscape, with a smoking chimney and a fragmented and repeated grid in different scales that refers to the constitutive elements of the electric pylons.

On the one hand, it is the calm and picturesque effect of an infrastructure that already seems to be obsolete and assimilated, that almost merges with the landscape; and on the other hand, the agitated fragmentation of an image that evokes movement and technical progress in a typically cubist geometric fragmentation. Through the expression of these two different approaches, one can distinguish a gap between the American and European avant-garde.

Lewis Mumford optimistically believed in this technological turn and thought about urban development through his regionalist theories.[10] He considered the telecommunications and transport infrastructures to be the keys to urban expansion. The automobile offered freedom of movement by allowing users to work and live in different locations. Similarly, the telephone allowed long-distance connections with individuals located elsewhere. For Mumford, these new technologies, combined with sensible urban planning, could guarantee improved living conditions, welfare, and social balance. But while the propaganda created by telecommunications engineering celebrated how the telegraph as well as the telephone represented revolutionary systems capable of contracting space, Mumford reversed this argument.[11] For him, telecommunications was part of a spatial distension effect through an extended system of networks. Urban concentration was therefore no longer a model for economic production, and the neighborhood scale ceased to be the ideal condition for communication between individuals. Mumford's idea was even more radical than the advertising rhetoric of telecommunications, because social and economic patterns were transformed by the introduction of innovative technologies. For Mumford, the notion of space is real and empirical, in contrast to a discourse prevalent in media theory and based on a figurative image of space. The radio, for example, allowed the distribution of audible information, such as sound clouds, within a real, circumscribed space, be it a private living room or a moving car.[12] In the regionalist urban model, radio stations, like universities and museums, would become real centers of local and autonomous culture, specific to a particular city, and belonging to a system that offered several options for social life.[13]

Unlike other technologies belonging to the factory model, for example, radio had increased American imaginative power around technological potential—as being capable of building a better future, even a perfect one. Radio technology, says Susan Douglas, evoked many fantastic prophecies, such as the extreme possibility of connecting the inhabitants of planet Earth with those of Mars. In 1898, at the beginning of the first applications of wireless telephony, the chief engineer of the British Post Office, William Preece, envisaged such a possibility.[14] But in the craze for radio technology, the American public had to be able to imagine the future. In the general enthusiasm of the beginning of the last

Hugo Gernsback with his isolator (external noise eliminator). Published in *Science and Invention* 13, no. 3 (July 1925).

Hugo Gernsback, illustration in *Radio for All* (Philadelphia / London: J. B. Lippincott Company, 1922): front page.

century, the transition from science fiction to an imaginable and real future came to be considered possible.

The Luxembourgish-American inventor Hugo Gernsback, editor since 1926 of the science fiction magazine *Amazing Stories*, imagined an ideal city under the regime of "radiocracy." His fascination with the power of radio developed in various ways as soon

Hugo Gernsback, *Radiocracy*, Christmas and New Year's brochure, 1944. George H. Clark Radioana Collection, Archives Center, National Museum of American History, Smithsonian Institution.

as wireless technology became available. As early as 1922, he published an essay, a guide for use and public information, entitled *Radio for All*.[15] Later, in 1944, he envisioned a society based on the power of radio—a radiocracy—which he described in a short story published in a Christmas and New Year brochure, during the dark years of World War II.

Coming from the Greek "Kratos" for power, one deduces that as democracy refers to the power of the people, radiocracy refers to the power embodied in or through the radio. For Gernsback, radio could become a unifying political force that would lead to a peaceful balance between nations, and the technology embodied a true social utopia, a reservoir of hope in a context of total disarray: "Already radio, through broadcasting, has broken down old boundaries and imaginary geographical lines. Soon, with actual power added, radio energy may be taken out of the skies for the use of all mankind. . . . When this is accomplished, humanity will rise higher than ever before, while the waging of wars will be made much more difficult. Radio-electronic power will make countries more dependent upon each other than ever. It will give all peoples of the world unprecedented prosperity, consequently they will hesitate long to make war upon each other."[16]

Gernsback was fascinated by the ubiquity of radio waves, with their invisible and pervasive power, which he imagined would one day be able to animate every kind of object. From cooking machines to flying cars—called radiotronicars—to antitheft devices and keyhole sights, radio was to provide energy for many revolutionary instruments. In the inventor's imaginary, this new source of power would thus create universal harmony and protect individuals from injustice by preventing any future war and destruction. Radiocracy was a totalizing vision of the future: with radio power becoming a unifying agent between individuals and nations.

This science-fictional tradition, which draws on an emblematic incorporation of modern infrastructures, would continue in literature. It is an ongoing relationship to time that gives these objects a special status in the landscape, as "the last traces of an epoch in which time still exists," in contrast to the temporal flatness of hypermodernity that characterizes postmodern society.[17] This "heroic" vision of progress, which is unmistakably imprinted in the very matter of technical objects and modern infrastructures, remains a temporal reference point and "a gas pedal of fiction."[18] For some, this tension toward the future is inscribed in modern infrastructure which "speaks to the future," to the point that it becomes impossible to offer a representation of the future without finding at some moment of the narration "an infrastructural device with the capacity to carry the imagination beyond the common measure."[19] The end of this craze for futuristic projection during

Hugo Gernsback, *Radiocracy*, Christmas and New Year's brochure, 1944, 5. George H. Clark Radioana Collection, Archives Center, National Museum of American History, Smithsonian Institution.

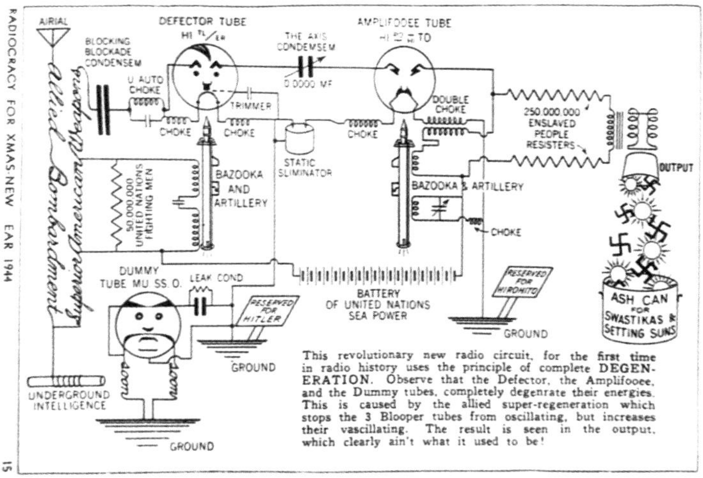

Hugo Gernsback, *Radiocracy*, Christmas and New Year's brochure, 1944, 15. George H. Clark Radioana Collection, Archives Center, National Museum of American History, Smithsonian Institution.

the postmodern period can also be observed and related to the environmental and economic crises on a global scale that marked society at the end of the thirty glorious years to the present day.[20]

Returning to Gernsback and the time when technical invention was still associated with the ideal of progress, we discover that these ideas were not foreign to Mumford. As a high school student in 1911, Mumford published his first articles in an issue of *Modern Electrics*, the first popular American newspaper on electricity, and edited by Gernsback himself.[21] But electricity was not the only thing that was important to Mumford. The world he imagined went way beyond any kind of science-fictional utopia: it was a vision rooted in the existing socio-economic and historical context instead. In 1926, when he was actively involved in the Regional Planning Association of America, Mumford contributed to the community planning experiments developed during the 1920s and 1930s: Radburn in New Jersey and Sunnyside in New York.[22] For Mumford, Regionalist principles were the basis for the development of a community plan. He and his colleagues considered them necessary steps toward reforming capitalist concentration and its degenerate urban model.

Although utopian ideals played a central role in his writings, Mumford kept his distance from any form of utopianism and foreswore any membership in the tradition of utopian thought.[23] His first book of 1922, *The Story of Utopias*, revealed his initial interest in the human condition and its ideal organization. However, Mumford denounced the utopian national model as a centralized form of propaganda that reflected the degeneration of the modern megalopolis, as embodied by the Dickensian city of Coketown: "Whereas the inhabitants of the national utopia may originally have been as diverse as the trees in a forest, they tend to become, under the influence of education and propaganda, as similar as telegraph poles along a road."[24] This image of standardization of the individual is identified with the telegraph pole and considered a consequence of the excessive centralization and idealization of the industrial society—without taking into consideration its history, its needs, and its circumstances. It is also the basis of Mumford's explicit criticism of *Looking Backward*, Bellamy's novel that anticipated the totalitarian setbacks that history was soon to experience.[25]

Toward the end of the book, however, Mumford introduces the term "eutopia," once again explained as a better society achievable in the present. He believed in the improvement of society "through sound design in architecture, practical reforms in housing, and comprehensive planning on the regional level."[26] In fact, the use of the term "eutopia" can already be found in Patrick Geddes's *Cities in Evolution*, a seminal text, as we have seen, for the understanding of regionalist principles.[27] The term "eutopia" is used in the following text: "the creation, city by city, region by region, of its Eutopia, each a place of effective health and well-being, even of glorious and in its way unprecedented beauty,"[28] which is opposed to the notion of kakotopia (a bad place) associated with the period of early industrial development. These moral meanings, the good and the bad, as well as the beautiful and the ugly, expose the discourse of urban planning to elevated levels of judgment well beyond the criteria of the form and proper functioning of cities. Urban construction and planning, for Mumford, could not lead to a truly better quality of life unless accompanied by shared ideals among individuals. The notions of urban decentralization and regionalism were based on the need to assign value to each individual—the inhabitant of the architecture, the city, the region, and even the nation.[29] Regionalism was an agent of social cohesion that, through modern technologies, allowed for emancipation from urban concentration and inhumane living conditions.

In another way, Wright also proposed an idealization of the American suburb, based on a positive use of modern technologies. Like the members of the RPAA, Wright believed that the contemporary urban landscape, a direct product of industrialization, was a dangerous consequence of the capitalist regime, and that a new type of urban planning would have built a healthier society based on more democratic principles. As we have seen, from 1932 to 1958, Wright worked on Broadacre City, an urban vision based on the idea that decentralization would bring society together around a project for a new community plan. One acre of land per person was the basis for a redistribution of goods and capital.[30] The one-acre-per-person principle would thus establish an extensive grid of urbanization connected by individual transportation and powerful communication networks. The rapid mobility offered by the automobile defined the spatial distribution of Broadacre, a continuous and diffuse system across the nation. The encounter between nature and infrastructure on a very large scale, as imagined for Broadacre City, would define the entire American landscape. What Wright called "electrical intercommunication" would motivate personal exchanges between individuals. Radio, telephone, and telegraph would have contributed to the reconstitution of individual identities, in direct opposition to the deindividualization of the working class in the capitalist system. The American inhabitant of this idealized structure, with his or her cultural, political, and social needs, became the standard model for this new organization.[31] Public services and government would be personified by the very occupants of Broadacre.

Curiously, the multipassenger airplane—as it was developing in his time—was not part of this urban vision and was substituted by individual machines that for Wright inevitably represented the future of aviation: "aerators" capable of traveling vertically and in all directions through reversible rotors and radio control. The drawings of these futuristic shuttles show similarities with the machines imagined by Gernsback—the radiotronicar—while differing in their source of energy. Wright was sensitive to the sensational aspects of science fiction literature, but he also maintained a somewhat romantic attitude toward traditional values, such as the house and its ideal occupant. Indeed, the architect imagined the disappearance of almost all means of collective transport, in order to restore the citizen's freedom of movement in space and time. Only a collective fixed train, a monorail traveling at 220 miles per hour, remains in his project.[32]

We have seen how Mumford's and Wright's ideas expressed the spirit of a unified America and proposed a model for the political future of their country through their urban theories. In both cases, infrastructure was to serve as an instrument of democratization and a reaction to the capitalist concentration of power and wealth: infrastructure is thus at the heart of their urban eutopias. The prefix eu- highlights their insertion in a particular American context, with both projects relating to specific political, social, and economic conditions.[33] For these more or less imaginative ideas were not simply abstract projections: more than just urban plans, they were intended to respond to local realities.

At the same time in Europe, modernity was producing other types of utopian visions, particularly in the St. Simonian technocratic tradition that characterized the nineteenth century. According to these ideas, industrialization and modern science would restructure society with the involvement of every individual in collective work, leading to a system of "true equality." These socialist ideals differed from American eutopism, which was focused more on the free initiative of the individual. Despite these differences, they shared the hope that technological change would serve as the engine of social transformation. Contemporary with the American eutopian explorations, Le Corbusier also reacted to the uncontrolled growth of capitalist cities. Beginning in the 1920s with a

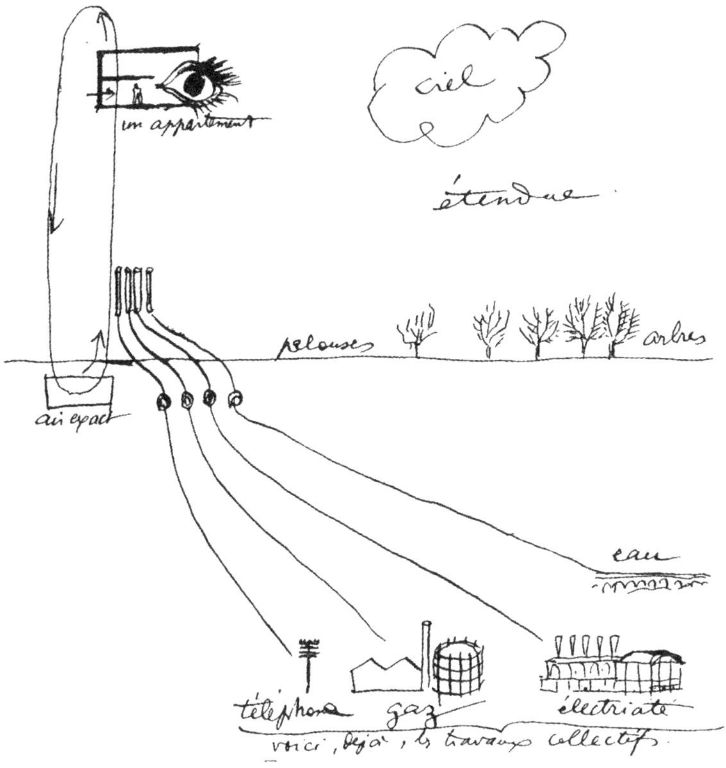

Le Corbusier, schema for the Radiant City from *La ville radieuse* (Paris: Editions Vincent Fréal &Cie, [1933] 1964), 36. © F.L.C. / 2023 Prolitteris, Zurich.

series of plans for the reconstruction of Paris and the 1922 project for a contemporary city of three million inhabitants, he would bring his thinking to fruition in *La ville radieuse*, published in 1933, a project for an ideal city for modern society.[34]

The Radiant City represented an urban model based on the exploitation of modern technology and infrastructure. While Mumford and Wright developed plans that were spread out and integrated into the natural landscape, Le Corbusier accentuated the process of urban concentration. The Radiant City expressed political and urban principles in the context of its time and the emergence of dictatorial regimes in Europe. The model of the Radiant City aimed to reconcile human civilization with nature through an organized distribution of glass and steel skyscrapers scattered in parks. Inhabitants lived in blocks—units—allocated on the basis of family size and need, not on economic status. Services were equitably shared by the population without any hierarchical order, so concentration became the most appropriate way to use these different common goods. The first public presentation of the Radiant City took place in 1930 at the International Congress for Modern Architecture (CIAM) in Brussels. Dedicated to the subject of garden cities, the Congress took as its starting point the decentralized model proposed by Ebenezer Howard, in opposition to the model of urban concentration. The garden city, which served as a source of inspiration for contemporary thinkers such as Mumford and Wright, became in Le Corbusier's eyes a form of social degeneration: "The garden city leads to individualism; in reality to a slave-like individualism, to a sterile isolation of the individual. It entails the destruction of the social spirit, the decline of the collective forces; it leads to the annihilation of collective will; materially, it opposes the fruitful application of scientific conquests and it limits comfort; it increases the waste of time, and it harms freedom."[35]

For Le Corbusier, urban concentration still carried a positive value by promoting the introduction of what he called "common services." These consist of the infrastructures that connect the inhabitants to the services of modern life: telephone, gas, electricity, and water. Embodied by horizontal lines that gather and rise vertically to the top of the residential tower, these infrastructures assume a primordial role, alongside a monolithic architecture, in one of his famous sketches illustrating the Radiant City. For Le Corbusier, the geometric abstraction of the line becomes almost a moral posture. However, in *The Poem of the Right Angle (Le Poème de l'angle droit)* (1955), it is by the figure of the right angle that he describes his action as an architect:

On a	With carbon
avec un charbon	we have
tracé l'angle droit	traced the right angle
le signe	the sign
Il est la réponse et le guide	It is the answer and the guide
le fait	the fact
une réponse	an answer
un choix	a choice
Il est simple et nu	It is simple and naked
mais saisissable	yet knowable

Les savants discuteront	The savants will talk
de la relativité de sa rigueur	of relativity and rigor
Mais la conscience	But conscience
en a fait un signe	makes it a sign
Il est la réponse et le guide	It is the answer and the guide
le fait	the fact
ma réponse	my answer
mon choix.	my choice.[36]

The angle, the part of the plane at the intersection of two half-lines, becomes a manifesto, a declaration of intent of the modern man who builds, who transforms the world by following the right sign. Vertical concentration thus made it possible to use common networks while remaining connected to the natural landscape surrounding the urban blocks. In Le Corbusier's words: "I think, on the contrary, that the present densities of cities, which are 300, 400, even 600 (in overpopulated neighborhoods), must be increased to 1,000 by the prodigious resources of modern techniques. It is then that common services can be multiplied, bringing effective liberations within family life, liberations in place of domestic slavery."[37]

Unlike Albert Robida's illustrated dystopian novel *The Twentieth Century: Electric Life* (*Le vingtième siècle: La vie électrique*), of 1893, which is set in a still more or less historicist French city, very dense and obscured by an uncontrolled multiplication of electric wires, Le Corbusier reinvents the structure of the city from scratch.[38] Robida's *Scrambled District* (*Un quartier embrouillé*) suggests threatening dysfunction through the disordered mass of wires in the sky. The appearance of a multitude of flying machines of all kinds further indicates the idea of a near future (precisely the year 1952), marked by a very free circulation in both horizontal and vertical directions. The author also imagined the "telephonoscope," a wall-mounted flat screen capable of broadcasting the latest information, theatrical plays, and courses and conferences, an amalgam of projected inventions in the field of telecommunications. Conversely, the orderly and simplified composition of Le Corbusier's design expresses a sense of control and perfect mastery of the technical networks, naturally conferring a political meaning to the project.

While Le Corbusier believed that urban planning was crucial to the realization of his ideal society, it was not the only

UN QUARTIER EMBROUILLÉ.

Albert Robida, *Un quartier embrouillé* from *Le vingtième siècle: La vie électrique* (Paris: Librairie illustrée, 1893), 129. Bibliothèque nationale de France.

requirement. The architect imagined a political organization that would combine an ideal quest for personal freedom with principles of cooperation, of authoritarian aspects with syndicalist values. As Fishman noted, Le Corbusier opened *La ville radieuse* by asserting that social organization rests on "individual freedom." But on the other hand, his publication bears a dedication to "authority."[39] Indeed, the political structure of his model was based on a technocratic and hierarchical system; he believed that only the powers of strong government could usher in the "Age of Harmony." Le Corbusier was, at this point in his life, in search of a kind of benevolent totalitarianism based on a technocratic model that could govern society. This was linked to his belief in a universal syndicalism, stemming from the St. Simonian tradition.[40] The hopes he invested in authoritarian systems is also to be understood in relation to the particular political context of Europe in the 1930s, and Le Corbusier's desire to find a government capable of adopting and applying his urban plans in their entirety.

If one compares the way in which the telecommunications companies developed during the early twentieth century, with the aim of applying a universal system that resulted in the constitution of a true monopoly, Le Corbusier's orientation toward totalitarian structures capable of sustaining the progress achieved by modernity somehow becomes a matter of course and fits into that form of euphoric belief in powerful modern technological systems as a political tool that had already marked the historical avant-garde. The inspired images of Futurist or Constructivist architects and artists, such as Antonio Sant'Elia, Virgilio Marchi, or Yakov Chernikhov, show how the modern industrial city, with its new programs such as post offices and telegraph services and stations, produced new aesthetic possibilities, grandiose urban scenographies traced by aerial threads that irrigated the skies from the tops of massive buildings, while also releasing a sense of power and control on a large scale.[41] These techno-futuristic visions would converge with the emergence of political regimes, respectively Italian and Soviet, that embodied all the ambiguity of this type of centralized representation of power.

Behind a complicated relationship to political power, the Radiant City represents an alternative society characteristic of its time. By adopting a model very different from those of Mumford and Wright, Le Corbusier promoted the idea of the harmonious organization of an ideal society, based on the functioning of modern technologies, or "common services," a technocratic vision that was to reflect the political order and consolidate its social structure. Then, how can we consider the belief in telecommunications systems and their operation as forms of utopian projects? Were Mumford, Wright, and Le Corbusier somehow dependent on Alexander Graham Bell's notion of eutopia, a "good place" that would be realized not only through the extraordinary

Antonio Sant'Elia, *La centrale elettrica*, 1914. Private collection.

progress of telecommunications, but also through the materialization of this progress in architecture and, therefore, in the form of infrastructure?

For Wright and Le Corbusier, the first step was to establish an ideal urban organization—one whose form would have pushed

Virgilio Marchi, Palazzo Postelegrafico (Project for Architecture), 1919.

forward and allowed a new social contract. Broadacre City was a kind of extension of the Jeffersonian ideal, each lot expressing the primacy of the individual, the citizen. Le Corbusier's plan was the result of a technocratic mastery of space coming from his St. Simonian culture. Mumford and the RPAA, on the other hand, did not propose an actual urban entity; their vision was on a larger scale: imagining infrastructures that accelerated the process of decentralization of urban populations into fractions, villages,

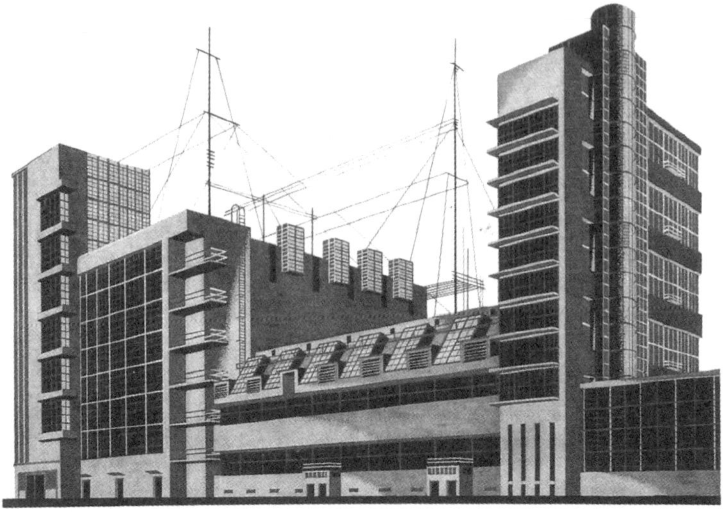

Yakov Chernikhov, *Composition n. 30* from *Architectural Fantasies: 101 Compositions in Color, 101 Architectural Miniatures* (Leningrad: Mezhdunarodnaia kniga, 1933).

towns, and hamlets—all connected to local resources by energy networks. What unites these three thinkers is the belief in a technology capable of creating true cohesion between people. Their respective critiques of the existing context led to contrasting models. These not only reflect an ideal organization of space, but also indicate how technologies, such as the telephone, could play a decisive role in empowering future inhabitants of cities and regions in an egalitarian way.

At the same time, the urban form illustrates how these three planners thought about infrastructural organization. Regional systems for Lewis Mumford, free and individual communications for Frank Lloyd Wright, shared services for Le Corbusier: these three models give shape to different urban distributions. If the spatial visions proposed by Mumford, Wright, and Le Corbusier were utopian, it was mainly through a combined conception of architecture, urbanism, infrastructure, and technology. This combination of material, technical, and informational scale represented an attempt to see human occupation as a "totality," a form of union to be shared and achieved for a higher good. They are thus three eutopic projects, profoundly modernist, precisely because they envisaged the creation of a unified political body. However, in reaction to the abstraction of the real geography imposed by the progressive technical regime and its capitalist exploitation, these proposals express the desire to remain connected with the resources of the territory.

They are alternative visions to what has been called "the loss of place," a constant and common prejudice of the great technical systems (railroads, electricity, telephone, oil) that have emerged from modern capitalism and that designate a progressive detachment of infrastructures from energy localisms, and the associated economies.[42] This general conduct established since the end of the nineteenth century would then have shaped a radical disjunction between the architectural and urban fabric and the environment, and will thus be challenged in a precursory manner by regionalist theories, and other eutopic visions of hyper-concentration, followed later by ecological, bioclimatic, contextualist approaches. In this sense, these dissident positions in the face of dominant interpretations of modern capitalism remain instructive in light of the contemporary phenomenon of splintering of cities and public spaces. This process is the result of a diffuse movement toward the privatization and liberalization of infrastructure. As already stated by geographers Stephen Graham and Simon Marvin, we are witnessing today a splitting of infrastructures, the creation of "high value" infrastructures, in a kind of scenario of urban life that is occurring à la carte.[43] The infrastructures, then, from their initial installation to the present day seem to embody different political models, starting with an initial ideological role as unifying agents, passing through a commercial and standardized application of the great technical systems and their services, up to a contemporary liberal perception that insists rather on the social fragmentation of its uses. The alternative visions studied here, beyond their value as urban projects, are based on social ideals that challenge the dominant economic and political models developed during the last century. Despite Bell's premonition, technology does not immediately offer a better place to live, but rather an interpretive framework for understanding the political, economic, and social context that in turn operates its deployment.

5 *Wireless: Theories of the Invisible*

While the wired infrastructure refers us to the design of a new modern landscape, other technologies have also contributed to accelerate and perfect our means of remote communication. As we saw earlier, the option of a world connected by an invisible, hidden, immaterial infrastructure has been the basis and inspiration for several projects and urban imaginaries. It is characteristic of infrastructure, and of technical systems in general, to be "on and off," at times invisible, hidden and camouflaged (hence *infra-*), and at other times visible and on display as manifestations of civic wealth and prestige, like the clock tower in medieval times.[1] Communication by hertzian waves—radio—followed two decades after the telephone first established itself as an alternative to the telegraph. In 1906, the simple signal of Morse code became voice, sound, and music, thanks to the invention of the amplitude modulator by Reginald Fessenden. From that moment on, radio became a platform for global broadcasts, from concerts to information, along with radio plays and several political and propaganda speeches that have marked moments in popular culture and history.

On October 30, 1938, George Orson Welles broadcasted a radio adaptation of *The War of the Worlds*, the story of an extraterrestrial attack in progress, which is said to have caused a collective moment of panic; on August 9, 1945, President Truman announced

The great antenna of the Eiffel Tower. Chromo published by the Chocolate of Aiguebelle company, about 1930.

to the American people the launching of the atomic bomb on Hiroshima; on August 28, 1963, Martin Luther King, Jr.'s famous speech "I Have a Dream" was broadcast live by radio and marked a decisive turning point in the American movement for civil rights. The playwright Bertolt Brecht saw the radio as the basis of an interactive communication network. With a single device replacing the theater, the lecture, and the newspaper, he imagined a network capable of making people talk and listen, of creating relationships to take the place of isolation, thanks to a "vast network of conduits."[2] It was a visionary prophecy that would soon become reality with the advent of the Internet.

Despite interpretations that associate a presumed dematerialization of infrastructure with recent (digital) developments in telecommunications, wireless technology has followed a history that parallels that of wired technology. The former did not follow upon the latter chronologically, and it too goes back to the end of the nineteenth century, with its own distinct uses and objectives. The absence of wire and the propagation of impalpable waves have largely contributed from the start to a vision of radio as an invisible technology. Beginning in the 1920s, several media theorists, from Brecht to Theodor Adorno and Rudolf Arnheim, insisted on the wireless and intangible nature of the technical medium that, coupled with its ubiquitous character, ended up erasing in the common imagination the architecture and objects of a network that nonetheless remained largely material.[3]

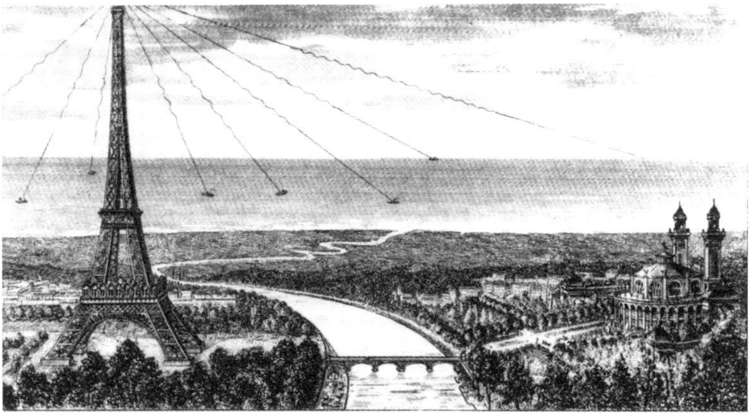

The Eiffel Tower sends dispatches by electric waves to America and to ships making the crossing within a radius of 6,000 kilometers. Original postcard text translation.

Yvonne Printemps's concert at the Eiffel Tower by TSF (*telegraphie sans fil*, or wireless telegraphy), 1922. Agence Rol. Bibliothèque nationale de France.

The intent of this chapter is to demonstrate that even wireless communication, supposedly aerial and immaterial, has a real and objective corporeality—comprising stations and antennas, territories and distances covered by established networks—that conveys precise intentions and projects. In fact, the invention in 1896 of the radio, used mainly for the transmission of information and programs at a distance, follows a path of implementation and

Aleksander Mikhailovich Rodchenko, *Radio Tower*, 1929. © 2023, ProLitteris, Zurich.

infrastructural development parallel to the telephone's. Based on a diffusion by electromagnetic waves, and invisible to the eyes, the radio infrastructure is often materialized by its majestic antennas and stations, which allow the radiation of information. The fascination for these technical objects was obvious from the outset.

Radio would save the Eiffel Tower from its planned demolition. In 1898, the engineer Eugène Ducretet demonstrated the possibilities of a Morse code telegraph link with the Pantheon, two and a half miles away, which was subsequently extended to London in 1899. Gustave Eiffel understood the interest of such an operation. In 1903, he financed the installation of an antenna support at the top of the tower and proposed to the pioneer of wireless telegraphy, the Captain of Engineers Gustave Ferrié, that he carry out a series of experiments. Ferrié then stretched six antenna wires from the top of the tower to the Champ de Mars. In order to muffle the crackling of the transmitter, the radio station was buried under

László Moholy-Nagy, *Berlin, Radio Tower*, 1928. Thomas Walther Collection. Gift of Thomas Walther. Museum of Modern Art (MoMA), New York, USA. © 2025. Digital image, The Museum of Modern Art, New York / Scala, Florence.

the Champ de Mars. As early as 1910, the French Army successfully transmitted signals to the Navy. The tower played a decisive role, as both transmitter and receiver, during World War I and managed to intercept several German messages.

Transferred from the Ministry of War to the Ministry of Telecommunications, Radio Tour Eiffel became the first radio station in France in 1921, a state broadcasting station that operated until

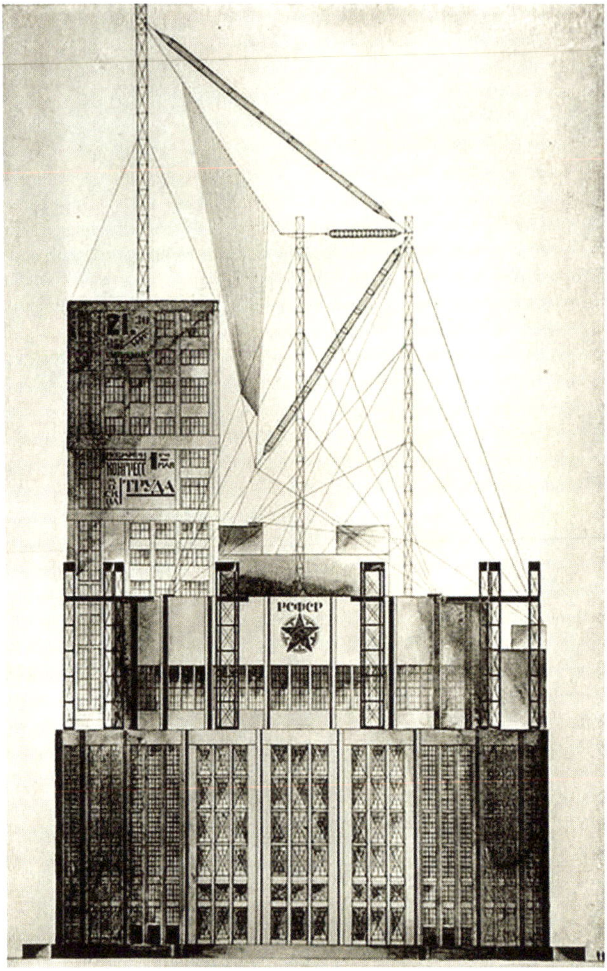

Alexander, Viktor, and Leonid Vesnin, project for the Palace of Labor, Moscow, 1923. *L'Architecture Vivante*, 1926. © 2023, ProLitteris, Zurich.

1940. From a symbolic object showing the prowess of modern engineering, the Eiffel Tower became an effective support for the broadcasting of French wireless telegraphy. Often associated with or compared by its vertical projection and reticular structure to the structure of the antenna, the Eiffel Tower came to integrate this very function for some forty years.

The object itself did not fail to inspire designers and architects. Vladimir Tatlin, who had visited the Eiffel Tower in 1914, imagined his monument to the Third International as an information center for the proletariat, a socialist planetary network, equipped not only with a newspaper and a telegraph, but also with a radio station; the tower was topped by two very thin transmission masts.[4]

Ivan Leonidov, project for the Lenin Institute, 1927. Courtesy of the Ministero della Cultura—Biblioteca di Archeologia e Storia dell'Arte.

Illustration by Le Corbusier, *Vers une architecture*, 1923.

The famous photographs of transmission towers taken by Alexander Rodchenko and László Moholy-Nagy during the 1920s similarly bear witness to an aesthetic infatuation with the technical object that effectively embodies the properties of wireless.

This "wireless avant-garde," as Mark Wigley calls it, characterizes a series of revolutionary projects that, by the presence of cables and reticular structures, transformed architecture into emblematic signs of connectivity, such as the 1926 competition project for the Palace of Labor by the Vesnin brothers, the project for the Lenin Institute by Ivan Leonidov in 1927, or the monument to Christopher Columbus, by Leonidov as well, from 1929.[5]

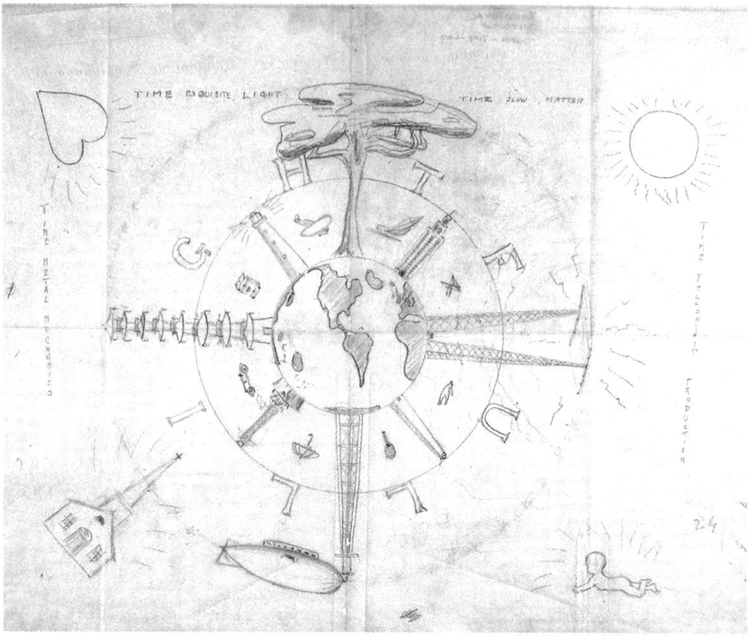

Richard Buckminster Fuller, manifesto/diagram of the "Lightful Houses," 1928. Buckminster Fuller Papers, Stanford University. Courtesy, The Estate of R. Buckminster Fuller.

More sensitive to the linear materiality of the telephone infrastructure as part of the set of common services connected to architecture through the articulation of wires, sheaths, and networks, Le Corbusier had a precise vision of the symbolic value of the line, as we have seen in the Radiant City. Moreover, his famous montage first published in 1921 in the *Esprit Nouveau* juxtaposed the greatest monuments of Paris with the silhouette of a contemporary ocean liner, equipped with two radio transmission masts. These are connected by a wire that borders all the monuments, to emphasize the technical achievements of his era.[6]

Buckminster Fuller, in the eyes of Wigley, recognized the absolute conditioning of the human being and his environment by the invention of the radio. As a result, Fuller was the architect who *par excellence* built *with*, *for*, and *in* the invisible universe of the electromagnetic waves. He presented himself through a symbolic autobiography: born in the year of Marconi's invention, serving the U.S. Navy during World War I as a radio operator, and witnessing the first transatlantic wireless communications. Fuller's work explicitly manifests the way in which this technology feeds

Illustration from *Collier's* (March 16, 1929). George H. Clark Radioana Collection, Archives Center, National Museum of American History, Smithsonian Institution.

a technicist vision of architecture. A series of photographs of radio masts were published in *Shelter* magazine in November 1923, and like electric pylons, they became manifest structures that symbolized his motto of "doing the most with the least."[7] For Fuller, radio equipment such as antennae would then become housing sites, as in the 4D Tower (1928), and any new house proposal, such as

Masts for U.S. Naval Radio Station, Dutch Harbor, Alaska. Photo by George Hanscom, 1915. George H. Clark Radioana Collection, Archives Center, National Museum of American History, Smithsonian Institution.

Triple T antenna at South Schenectady Laboratory, n.d. George H. Clark Radioana Collection, Archives Center, National Museum of American History, Smithsonian Institution.

Radio antennas, R.C.A. Communications Inc., New Brunswick, New Jersey, n.d. George H. Clark Radioana Collection, Archives Center, National Museum of American History, Smithsonian Institution.

Tesla Central Power Plant and Transmitting Tower for "World Telegraphy," Long Island, New York, n.d. George H. Clark Radioana Collection, Archives Center, National Museum of American History, Smithsonian Institution.

Coil connection house with metal insulators, Tuckerton, New Jersey, n.d. George H. Clark Radioana Collection, Archives Center, National Museum of American History, Smithsonian Institution.

Antenna and station, New York Navy Yard, 1907. George H. Clark Radioana Collection, Archives Center, National Museum of American History, Smithsonian Institution.

the Dymaxion House (1929–1930), systematically supported a radio mast—to the point of assimilating its structure. The 1928 "Lightful Houses" diagram is absolutely explicit in its reduction of the planet to a globe nested within a larger sphere of radio communication. Transmission masts, a mooring point for airships, and a large tree define the circumference of the system, which includes other signs of modern construction: vehicles, objects, leisure activities, buildings. Beyond it are the sun, a heart, a church, life. Fuller's architecture clearly occupies a space of communication rather than of earthly contingency; it is situated in another dimension, "rather in the air than on the ground."[8]

Although reticular radio towers and antennas tend to symbolically overshadow all the equipment and architecture required to produce radio broadcasts,[9] and clearly fascinated a generation of architects and designers, they also bear witness to a highly artisanal, authorless production process. Especially in the early days, they are endlessly tinkered with, manipulated by inventors and entrepreneurs, but also by private users. They are thin, high, wide, cylindrical, mushroom-shaped or diamond-shaped; they are majestically placed on the ground, in pairs, in threes, or in isolation; they are portable, placed on ships, boats, planes, cars. The towers and masts with their connecting wires differ from the telephone line in their open field of possibility, their expansive composition, and their expressivity.

From the outset, wireless or underground technologies represented costly alternatives for remote communication, compared to the "classical wired landscape." The choice between these options originally conformed to a precise political design, as in the project for the Imperial Wireless Chain, proposed in 1910 by Marconi's Wireless Telegraph Company for a wireless telegraphy network connecting the countries of the British Empire, which was completed by the State Post Office company in 1928.[10] From the beginning of their installation, wired or wireless technologies subtended two different political projects. They made an explicit contrast between a vision of communication as a national resource, conforming to the physical boundaries of the territory and enabling unity through an internal system of networks, or alternatively, as a colonialist conception of communication that would pass through the electromagnetic waves present in the atmosphere and bypass geographical borders between countries. At that time, these two projects—the affirmation of national cohesion or of colonial influence—were two facets of a political ambition common to several nations: total and dominating power on a world scale.

The figure of Guglielmo Marconi himself combined both the genius of the inventor and the vision of a great entrepreneur, accompanied by a strong nationalistic pride. This would lead him to assume real political responsibilities within the Italian government of the time and perfectly expresses the ambiguity of the progressive spirit of the modern era, in a historical and political context strongly marked by the emergence of totalitarian regimes.[11]

In *Empire and Communications*, Harold Innis proposed a reading of history that highlights the way in which different media are used to communicate influence and the emergence and decadence of empires. These were encoded in stone, clay, papyrus, and the alphabet for Egypt and Babylon; by the oral tradition for ancient Greece; the written word for the Roman Empire; through the influence of parchment and paper for Europe in the Middle Ages; up to the printing press for modern times.[12] Following this line of thinking, for Friedrich Kittler, the media are situated above our lives in an even more radical way. They determine our position in relation to the world: the gramophone analogically reproduces the real by replicating the voice, the film displays the imaginary by capturing light effects, and the typewriter traces the symbolic by fixing written language.[13] These establish a true separation between sound, image, and language that, for the theorist, conditioned the course of the recent history of our civilization. Without overstating a deterministic interpretation of technological progress, one can observe the way in which the form and the material organization of infrastructures, whether wireless or wired, can also carry political meanings.[14] Technology is not, therefore, an autonomous agent that intervenes on society by forging it unconditionally, but its evolution and constitution can be read and interpreted in order to understand the environment in which it develops.

The mode of transmission of radio waves best represents what Marshall McLuhan called “acoustic space,” a term that originated in the multidisciplinary exchanges of the Culture and Communication Seminar held between 1954 and 1955 at the University of Toronto: an important moment of intellectual debate between representatives of media theory, architecture, and urbanism, embodied by McLuhan, Sigfried Giedion, and Jaqueline Tyrwhitt.[15]

Adapting the concept of "auditory space" introduced by the psychologist Carleton Williams, McLuhan applied this acoustic metaphor to the world of electric media, producing a space characterized by the simultaneity of sensory stimulations, whose central subject is everywhere and whose limit is nowhere.[16] Electric information, in McLuhan's interpretation, contrasts with the linearity of the book and the flatness of the image, creating a return to invisible and multidirectional space. Through the electrification of writing, beginning with the telegraph, script is transformed into sound, marking a big step toward nonvisual and audible space. Further steps, he stated, would then be taken by the telephone, radio, and television:

> The telephone: speech without walls.
> The phonograph: music hall without walls.
> The photograph: museum without walls.
> The electric light: space without walls.
> The movie, radio and TV: classroom without walls.[17]

Through the immaterial nature of radio waves or other properties of modern media, architecture metaphorically dissolves into a space without walls, without matter, becoming a place defined only by the activities that are carried out there. The acoustic metaphor used to express the idea of wave propagation is not new. In his work of popularization, *Radio for All*, Hugo Gernsback used the image of a stone thrown into water—the same image used by Vitruvius to visualize the sound wave—to describe the multidirectionality and concentric progressions of the radio phenomenon.[18] As we know from experience, the more powerful the sound source, the further the signal will be clearly heard; in the same way, the signal of a radio wave will be picked up according to the strength of the relay. But much more powerful than the sound wave, as Gernsback specifies, the radio wave crosses solid matter, it goes "through solid stone walls," and it passes "through mountains." Like the sound wave, it travels through liquids and air. Its speed of propagation is the same as the speed of light, and much greater than that of sound. Finally, in this chain of analogies one can only note the way in which modern telecommunications and infrastructures called into question architecture and its spatial properties.

6 *Law: The Public and Private Spheres*

If, on the one hand, architecture is conventionally considered a fixed and static object, infrastructure, on the other hand, refers to a system that carries dynamic flows and elements. Determining whether the house comes before the road, or vice versa, is an impossible exercise that, as others have already discussed, only highlights the way we look at these objects.[1] Moreover, architecture and infrastructure share the prerogative of occupying space and providing specific functions. What happens, then, to the nature of this space once it is invested by systems of telecommunication and particularly by networks of the sound transmission? How do the objects that constitute architecture and infrastructure address the dimension of the private as inseparably linked to the public, when they are both part of a single operating circuit?

To address this figure of the circuit, this chapter begins with what exists outside the built environment (or rather what is generally considered the public domain), before touching on the proper architectural object, the interior domestic space. This physical progression also reflects a theoretical requirement that derives from the difficulty of talking about infrastructure systems and movement of services by isolating the architectural scale from the urban and territorial scales. As we shall see, circumscribing the public

Connecting outlet outside a house, n.d. George H. Clark Radioana Collection, Archives Center, National Museum of American History, Smithsonian Institution.

and private dimensions of space is a complex issue that cannot be limited to the definition of simple objects.

As other authors have pointed out, the notion of the private is first and foremost a question of law, and is therefore less universal than the idea of intimacy for example, which goes beyond a simple legal definition.[2] The term "intimate" derives from the Latin *intimus*, the superlative of interior, and designates that which is hidden from an external view.[3] If intimacy is by definition secret, invisible, and silent, then privacy can be visible, tangible, and

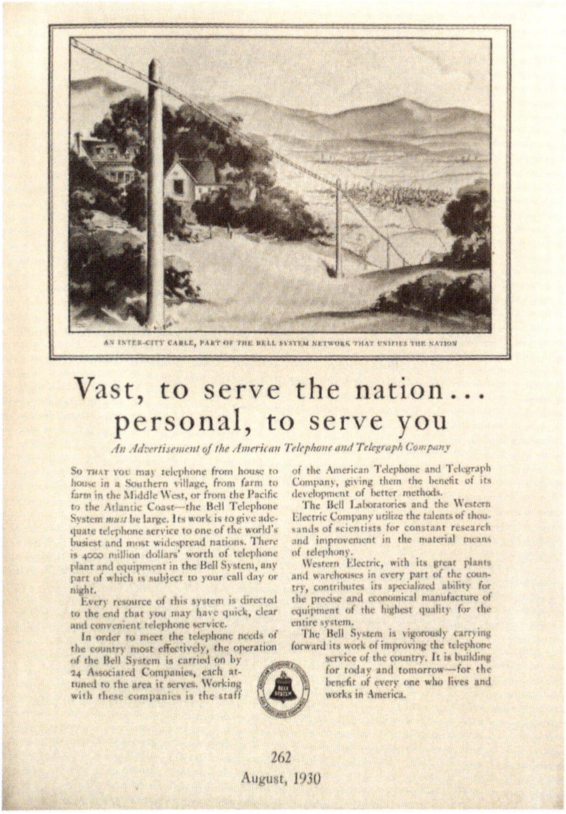

AT&T advertising, 1930. Courtesy of AT&T Archives and History Center.

audible. Yet it is perceived as something that must be protected from intrusion from the outside. The term "public," on the other hand, originally indicates that which belongs to and comes from the people.[4] Since ancient times, this authority—the public—has been represented by the state. The *res publica* thus designates what belongs to the community and what cannot be altered by the law. The distinction between the private and public is not a matter of physical location, but rather of how power is distributed.[5] Thus, the law becomes the real subject of this regulation.

From this perspective, we can then ask how modern telecommunications has developed in relation to the law. Was the expansion of sound networks across cities and territories a harmonious or a controversial process? And how did different cultural and political contexts, notably in North America and Europe, deal administratively with questions of private law and public use?

Western Union Building, 195 Broadway, New York, before the fire of 1890. Western Union Telegraph Company Records, Archives Center, National Museum of American History, Smithsonian Institution.

In the late nineteenth century, the lawyer Edward Q. Keasbey published *The Law of Electric Wires in Streets and Highways.* In this book, the author discussed a series of controversial cases related to the installation of networks and the use of transducer systems, which at that time were undergoing substantial expansion in the United States: "It has been found that the wires will carry the sound of the voice and send spoken as well as written messages. . . . For all these uses the electric current must be distributed from central sources and carried by wires throughout the city and across the country."[6] During the early years of telecommunications development, the number of poles and wires installed in the landscape increased considerably.

This new "occupation" of the roads made these objects problematic for property owners and landowners. The utility poles, Keasbey noted, disfigured the appearance of the streets and largely obstructed their use. The wires blocked access to buildings in case of fire and caused much damage during storms and floods. Some carried high electrical current and could become fatal to passersby, and in the case of wires with different applications intermingled, it was common for one current to interfere with another. On the one hand, these issues caused concern about the use of electric wires on roads and met with public opposition that sometimes led to actual litigation. On the other hand, for Keasbey wires had become indispensable because of their transmission function, which was now widely considered desirable and necessary. While legislation recognized an extension of the use of roads by the installation of a system of "transmission of intelligence"—the telegraph—it was also argued that transportation should be maintained as the primary function of the road, and that the multiplication of wires and poles could become a nuisance for circulation and for the private owners of adjoining land.[7] It was up to each municipality to adjudicate disputes by deciding what types of use were considered legitimate for any country or city road, and what types were not.

In an 1885 case in Saint Louis, Missouri, the Julia Building Association brought an action against the Bell Telephone Company following an injunction filed by the owner of a shop to prevent the telephone company from cutting a hole in the pavement in front of his building and installing a pole, as well as drilling a hole in the stone wall supporting a vault under the pavement. The plaintiff owned the right of way up to the middle of the street, while the telephone company was authorized by the legislator and the municipal council to build the line. The court found that the roadworks related to the telephone services were nevertheless necessary for a new use of the street that was not disruptive to the old one and was therefore appropriate for a city street.[8] Opposing counsel appealed, based on the argument that "the telephone does not need the street for its construction, and there is no connection between public travel and talking through a telephone." The final decision further confirmed that if the telephone was considered legitimate, there was no need to ask permission because each individual or company could make legitimate use of the street without asking permission from the city. Other cases proceeded along the same lines, with interesting nuances concerning the sources of the nuisance. In a dispute the following year in Washington, D.C., between a private citizen and the Western Union Telegraph Company, which had previously purchased the land in question, the clutter of poles, the source of danger constituted by the wires, and the annoying whistling of the wind through them were claimed to be detrimental to the primary use of the street.[9] In another case from 1898, in New Jersey, this time concerning the running of wires over a portion of private land, and not the installation of a pole, the local legislation specified that the running of wires did not constitute a violation of the private right of the land, unlike the installation of a pole.[10] A long series of other cases supported the opposite arguments, recognizing the right of land as belonging to private individuals and distinguishing between the use of the street as such and its use as a site for the installation of telecommunications networks. Cutting down trees to make room for poles and wires was another major subject of dispute, which oscillated between the idea of the legitimate use of the road and its role as a support for the telephone. Most often, when the companies lost their case, compensation was awarded to resolve the dispute, although sometimes the infrastructure was ordered to be dismantled.

Once again, Keasbey's conclusions involve links between the usefulness of telecommunications networks and the role of transport networks—or highways—by emphasizing that it was precisely because of the existence of highway infrastructures that the wires could fully perform their connective function: "their relations, therefore, to the public and private rights in the highways must be defined, conflicting rights must be adjusted, and the manner of using the street must be controlled."[11] As we have already seen, from its inception, telecommunication—and communication altogether—was not comprised of different autonomous systems (telegraph, telephone, and radio); rather, it was the result of a combination of networks and infrastructures that became dependent on each other, including transportation and the movement of flows in general. There was no highway or railroad traffic, for example, without telecommunications, electrical power and lighting sources, water, and gas; the reverse is also true. For the lawyer, the regulation of the streets was therefore a matter of both public and private law. "The streets are opened for public use," he said, "but

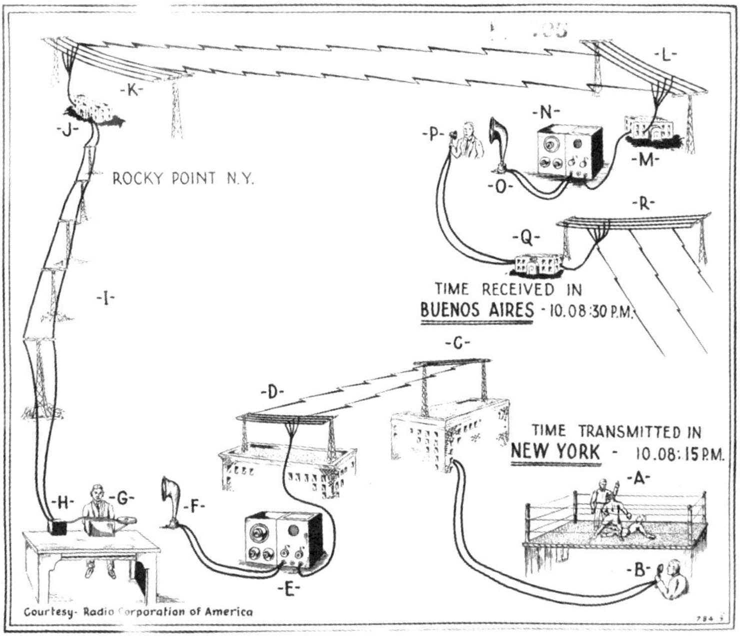

Diagram of the radio transmission circuit of the Radio Corporation of America, George H. Clark Radioana Collection, Archives Center, National Museum of American History, Smithsonian Institution.

only for use as a street."[12] Title to the land that was included in this use, however, remained in the name of the owner of the adjoining land. This discussion, in other words, questioned the true use of streets and the extent to which these traffic networks should accommodate other infrastructures that might interfere with that use, as in the case of electrical wires. Finally, it questioned the sphere of public and private law in relation to street use and other possible applications.

The North American development of sound transmission networks was obviously complicated in terms of legal definition. Although their service was considered a public good, these infrastructures were owned by private companies, and they occupied fragments of land, underground space, and underwater corridors administered by both private and public legislation. Keasbey thus identified a potential conflict that involved private infrastructure, public service, and private or public ownership of the land.

The framework and terms of the question were different in Europe, where the notion of public space traditionally plays a different role. In his brilliant essay "Knowledge of the city and the

Illustration of the *Pictorial Review*, 1922. George H. Clark Radioana Collection, Archives Center, National Museum of American History, Smithsonian Institution.

home in the early nineteenth century" ("Savoirs de la ville et de la maison au début du 19e siècle"), François Béguin used a series of English and French examples to explore the emergence of what he calls the "politics of housing." Through the study of various reports from the period, the author highlighted the way in which housing policy tended to be defined in terms of uses, functions, regulations, and services, rather than by fixed spatial or architectural categories. Water regulations, building and city cleaning services, and environmental comfort (air and light) were all new

Illustration, *Saturday Evening Post* (May 20, 1922). George H. Clark Radioana Collection, Archives Center, National Museum of American History, Smithsonian Institution.

elements produced by the sanitation apparatus of nineteenth-century housing policy.[13] In the early days, cleaning services were the most important element of housing policy. At the outset, the cleaning and distribution services of the housing networks were managed by the inhabitants, but these were gradually taken over by specialized professional figures. Civil engineers rather than architects were engaged in infrastructure projects to find solutions to the fatal problems of insalubrity.[14] In Europe, the first private companies in charge of various services were quickly taken over by the institutions of public administration. The management of

Radio listeners in a living room. George H. Clark Radioana Collection, Archives Center, National Museum of American History, Smithsonian Institution.

garbage, water and gas, and light—and later telecommunication—became primary concerns at the heart of a fully European conception of public space as a shared, state-owned good. Indeed, in 1889 in France, and in 1912 in the United Kingdom, telephone service was nationalized and thereby defined specifically as a public good.

Béguin concluded his study by describing a complex shift between public and private practices that occurred throughout the nineteenth century. Activities that were once public—that took place outside the home—were progressively privatized or domesticated through the appearance of new devices and services such as running water, electricity, telephones, radios, and finally television, inside the home.[15] These local systems depended in reality on entire urban and territorial networks that belonged to the public sphere and were set in a territory that was also very largely public.

The received knowledge of the house as a physical container defining the private sphere deserves to be rethought, as connections and other means of communication with the outside world began to break through the walls of what others have called "the private shell."[16] In visual culture, moreover, the boundary between the private and public sphere is traditionally defined—at least in the realm of domestic architecture—by reference to notions of interiority and exteriority. In other words, the difference between the public sphere and private sphere is defined by contrasting what happens inside or outside the walls of the house. The intimacy of domestic space thus corresponds to the ultimate idea of privacy, in contrast to the more heterogeneous notion of public space.

This established distinction has been challenged by a reading of architecture as a media object, which points to the modern era as a turning point. The use of the window in Le Corbusier's work is cited as a significant example of an architectural device that allows us to see outside the home, but also acts as a showcase to display certain elements of private life. Modern architecture would thus have opened a new concept of architecture by inaugurating what Beatriz Colomina calls the "publicity of the private."[17] If one considers as well other "atmospheric" properties of architecture such as light, sound, and climate, this distinction between interior and exterior becomes even more fragile. Walls and other physical barriers have never offered an absolute guarantee in terms of acoustic insulation, for example, since sound propagates through air and water, but also through solid matter. In their application to communication, therefore, modern technologies have strengthened these sound transmission properties to the point of completely overcoming material obstacles and being able to make the sound signal travel over unlimited distances, beyond any physical barrier. From the sound point of view, then, the limits between the public and private spheres have always been very blurred considering that sound is a dynamic phenomenon that travels from a source to a receiver (natural or technological), rather than a static object easily identifiable and spatially determined. Therefore, considering architecture and infrastructure as part of a continuous circuit forces us to move away from consideration of architecture as a purely formal object, and to think more in terms of powerful devices and mechanical functions.

If telecommunications allows the transmission of sound by means of various technologies and infrastructures, the latter become elements that are difficult to identify in terms of their private or public nature. The association of the house with the notion of the private requires being called into question again. From the Latin "*mansionem*," the accusative of *mansio*, which indicates the action of staying, along with all the activities that are carried out there, slips by metonymy into the house as a place

Wireless Telegraph Motor Car used by Marconi (at rear right) in 1901. Society of Wireless Pioneers / California Historical Radio Society, Alameda, California.

and as a physical space. The conception of the Roman or medieval house also implied a significant presence of the public.[18] The family home was often connected to the workshop or store, and the openness to the outside world was indispensable for a series of services necessary for everyday life. It was in the early seventeenth century, with the growth and enrichment of the bourgeois class, that the presence of the public inside the house began to diminish.[19] With the rise of the European bourgeoisie and the American middle class during the nineteenth century, the house became more and more private, with fewer and fewer exchanges with the outside world. The comings and goings of clients, employees, and servants were reduced. However, at the same time, there was an

Illustration on a set of cigarette cards with the Wireless Telegraph Motor Car used by Marconi, early twentieth century. Courtesy of Bart Lee, California Historical Radio Society, Alameda, CA, USA.

interest in the outside world, which resulted in the introduction of different forms of media, including books, maps, newspapers, and magazines. These media allowed contact with the outside world and were collected in specific rooms of the house, such as the office or the library. This way of connecting individually, or in small groups, with the public sphere continued significantly with the introduction of modern telecommunications.

Just think of the very confidential listening possible via the telephone or the way the radio system has evolved over time. In the beginning, the still-weak radio signals required headphones to hear them, but by the 1920s, when the system became widely available, and its technology substantially improved, listening to the radio became associated with moments of socialization between groups of people and families. From the automobile's audio system to the Walkman, to the various mp3 listening devices, the practice of listening to portable and personalized sound environments has become increasingly widespread.

These sound bubbles become private transportable entities outside one's own home. This phenomenon reflects a process of atomization of the modern society, where individuals in perpetual movement transport a part of their private sphere wherever they wish. Several authors, following Jane Jacobs's critical positions on postwar urban planning, speak of a privatization and virtualization of public space in the twentieth century through the multiplication of networks and the transformation of the individual from citizen to consumer.[20] If the infrastructures of the nineteenth century still offered public places of exchange and encounter within a dense and mixed city, the dispersion and atomization of the American suburban system would lead to the collapse of the public sphere, to the mobile private bubbles that partly characterize our current lifestyles. From these considerations, again, the spatial

Humorous vignette, n.d. Warshaw Collection of Business Americana, Archives Center, National Museum of American History, Smithsonian Institution.

definition of the private and the public seems to be determined not by architectural properties, but rather by everyday practices, regulations, and activities, which evolve over time. The domestic space can become private or public according to use, just as the public space can be crisscrossed by private environmental bubbles attached to the movement of individuals.

More generally, the history of telecommunications questions the definition of architecture and infrastructure in relation to its different practices. Keasbey's work on the initial development of infrastructures reveals the legal complexity that surrounds these networks. Architecture and infrastructure can hardly be isolated objects as long as one interferes with the other and one confers a function on the other. When we focus on uses rather than on objects, separating the architectural scale from the territorial scale also becomes an artificial operation. And, as we saw earlier, modern technologies such as the telegraph, telephone, and radio have emphasized the role of physical networks in connecting the private and public spheres. In other words, modernity "made habitat explicit."[21] The progressive technical specialization of networks has led to a separation between the functions imputed to the physical space of the house, and those that become legible through a "network mechanism": the presence of tubes, cables, sockets, and switches marking a possible contact with the outside world. This distinction of functions has been perceived as a prerogative of modernity.[22] As a consequence, one individual can communicate with another without physically moving, without seeing or feeling the presence of the other.[23]

This decisive and direct dissociation between seeing and hearing made its first explicit appearance in communication when on March 10, 1876, Alexander Graham Bell called his assistant Thomas Watson for the first time, pronouncing the phrase "Mr. Watson. Come here—I want to see you!" In telephoning from one room to another in the Boston Bell Laboratory and using only a few words, he was demonstrating the possibility of isolating different levels of communication—notably those of listening and seeing—by piercing "the shell of privacy" with an electric wire.

7 *Planning: Sound Conduits*

The invention of the telegraph, the telephone, the radio—and even the phonograph—is usually associated with ingenious characters and powerful narratives that emphasize how much these machines have changed our perception of sound and the very way we hear. "Mr. Watson. Come here—I want to see you!" said Alexander Graham Bell to his assistant in 1876, over the first telephone connection. Thomas Edison heard his own voice reciting a popular nursery rhyme—"Mary had a little lamb"—recorded on the cylinder of his first phonograph built in 1878. "Are you ready?" was the first wireless message transmitted in Morse code in 1897, by Guglielmo Marconi, who established his own company that same year. Some social and cultural revolutions—we are told—are linked to turning points that allowed sound to lose some of its ephemeral character and to bring people back in time or across long distances.

As Jonathan Sterne argues, these are powerful stories because, by fixing a moment and building a kind of myth, they tell us that something happened to the nature of sound at the end of the nineteenth century.[1] But these stories are also historically incomplete. They tend to erase the practices, technologies, and ideas that emerged before the inventions did. In our collective imagination, sound reproduction technologies, such as the telephone and radio, are closely linked to the invention of that ultimate device:

Charles Kuhn, "A chaque maison son téléphone," advertising, 1939. Museum of Communication, Bern.

the receiver or sound machine that allows an entire electrical or electromagnetic system to function.

Yet it is useful to go beyond the devices themselves and take a broader perspective. Technical progress is inscribed in a history of resistance of old technologies and not just in the forms of rupture created by new inventions, as some progressive narratives of history tend to describe them. By addressing systems and their use globally, "the daily life of techniques" allows us to better understand the complex evolution of a society, more than the separate study of each technical innovation.[2] The aim of this study is to explore the conduits that power the telephone and radio systems, to understand how they have become part of architectural practice through their integration into planning and design.

As Ithiel de Sola Pool observed, the telephone has contributed not only to the establishment of cities and the dispersal of activities in the territory, but also to the creation of new systems of living, from work to housing, by forging architectural typologies that accommodate them. The development of the skyscraper is based on several technologies. The elevator is commonly mentioned as the crucial device enabling its functionality, but for Pool, the telephone also played a decisive role, making it possible to replace the obsolete figure of the messenger boy within these architectures of density.[3] Today we could also formulate the opposite hypothesis by observing the way in which the telephone was able to develop thanks to a social and urban environment that was in the process of spatial reorganization, both horizontally (through the landscape) and vertically (through architecture) and extending to a large-scale operating structure. The telephone offered an effective solution to the primordial need to communicate within a new spatial regime.

This chapter will remain at the scale of architecture, particularly the scale of the house or even the apartment. The focus here is on the original documents produced by the American telecommunications companies at the height of their popularity, from the late 1920s to the 1940s, whose purpose was to address architects, engineers, and consumers. A 1928 American Telephone and Telegraph Company (AT&T) booklet, *Planning for Home Telephone Conveniences*, illustrates the growing popularity of telephone services and their installation in American homes. These increasingly required the development of innovative design strategies by various stakeholders in the manufacturing process: owners, architects, engineers, and builders. "The placing of facilities for telephone wires and

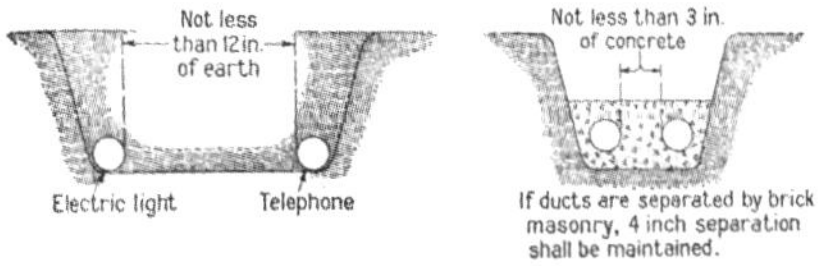

Illustration from American Telegraph and Telephone Company, *Planning for Home Telephone Conveniences: For the Use of Architects, Engineers, Builders, and Owners* (1928). Warshaw Collection of Business Americana, National Museum of American History, Smithsonian Institution.

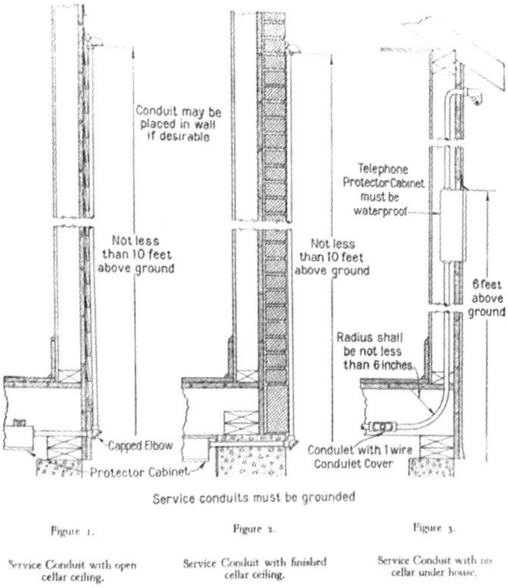

Illustration from American Telegraph and Telephone Company, *Planning for Home Telephone Conveniences: For the Use of Architects, Engineers, Builders, and Owners* (1928). Warshaw Collection of Business Americana, National Museum of American History, Smithsonian Institution.

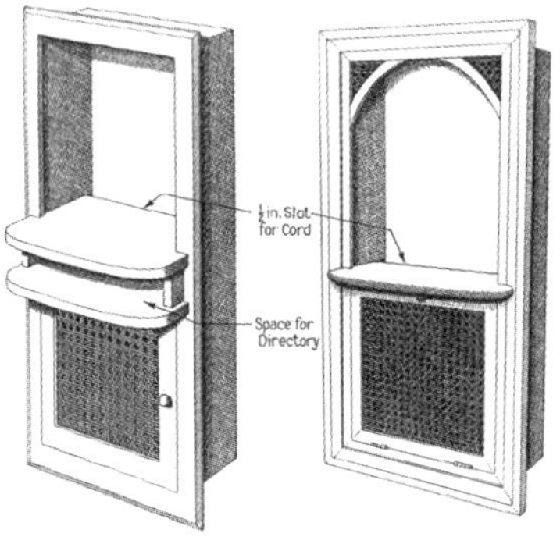

Illustration from American Telegraph and Telephone Company, *Planning for Home Telephone Conveniences: For the Use of Architects, Engineers, Builders, and Owners* (1928). Warshaw Collection of Business Americana, National Museum of American History, Smithsonian Institution.

Illustration from American Telegraph and Telephone Company, *Planning for Home Telephone Conveniences: For the Use of Architects, Engineers, Builders, and Owners* (1928). Warshaw Collection of Business Americana, National Museum of American History, Smithsonian Institution.

Illustration from American Telegraph and Telephone Company, *Planning for Home Telephone Conveniences: For the Use of Architects, Engineers, Builders, and Owners* (1928). Warshaw Collection of Business Americana, National Museum of American History, Smithsonian Institution.

apparatus during construction results in general improvement of the residence in that better appearance is secured by concealing the wires and some of the apparatus."[4] A new science of planning was developing for the integration of telephone equipment, whether concealed or visible, conceived for immediate or future installation. "Conduits, protector cabinets, telephone cabinets, or other arrangements for concealing the wires and apparatus," the booklet stated "must necessarily become a permanent part of the building."[5] In other words, the installation of telecommunications was no longer simply a matter for technicians and engineers. It also became an architectural concern.

The brochure goes on to list technical and functional details through a multitude of examples. Telephone conduits, wires, cables, switches, and sockets had to enter the home as unobtrusively as possible. Rigorously separated from the electrical wiring, telephone conduits usually penetrated buildings through the outer walls and rose from the ground up, through the cellar ceiling or the second floor of the building. More rarely, they came in from above, through the attic. Telephones in domestic interiors were usually installed in specially designed booths or on existing tables, supplemented with a built-in bell, assembled on site, or as a wall-mounted telephone handset. Comfortable benches, sofas, and armchairs were added to the installation of these devices to facilitate the integration of this new technical form of oral communication.

The generalized craze for sound reproduction technologies resulted in a surprising need to display these emblems of modern acousmatic sound ("acousmatic" being the common designation of a sound that one hears without seeing its original source).[6] As a material manifestation of invisible communication, the range of expressly designed furniture and objects underlined the existence of novel machines with the ability to reproduce an original sound that was far away and its source invisible. This distinction between seen and unseen objects, services, and infrastructures became the focal point of the entire AT&T publication. Following the principle of North American house kit catalogs (the most popular being the Sears modern house catalog), the booklet presents a range of solutions for different houses, family configurations, and needs.

From the single-family residence to the duplex or multi-unit building, from the modernist apartment to the vernacular or country house, the authors proposed a specific installation of telephone equipment to obtain the most functional, comfortable,

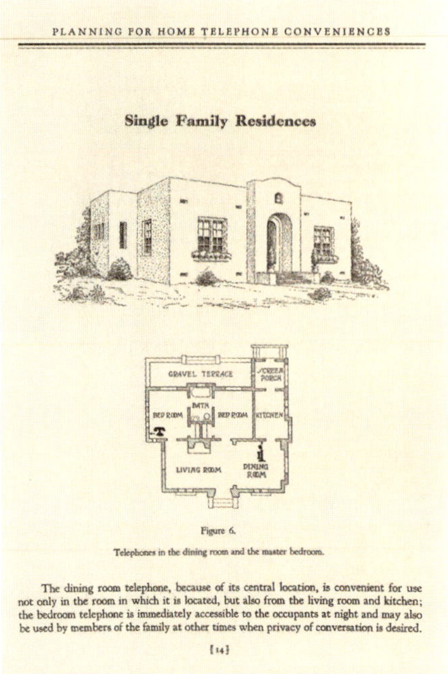
PLANNING FOR HOME TELEPHONE CONVENIENCES

Single Family Residences

Figure 6.

Telephones in the dining room and the master bedroom.

The dining room telephone, because of its central location, is convenient for use not only in the room in which it is located, but also from the living room and kitchen; the bedroom telephone is immediately accessible to the occupants at night and may also be used by members of the family at other times when privacy of conversation is desired.

[14]

Illustration from American Telegraph and Telephone Company, *Planning for Home Telephone Conveniences: For the Use of Architects, Engineers, Builders, and Owners* (1928). Warshaw Collection of Business Americana, National Museum of American History, Smithsonian Institution.

PLANNING FOR HOME TELEPHONE CONVENIENCES

Figure 7.

Telephones in the living room, the front bedroom and the kitchen. Additional telephone outlets in the rear bedroom and the dining room are suggested.

The living room telephone is in a convenient location for comfortable day and evening use; the bedroom telephone is readily available in case of sickness, alarm, or other emergency situations at night; the kitchen telephone permits prompt answering of incoming calls and is useful in the conduct of the household affairs, such as marketing.

The other telephone outlets are suggested in the interest of greater comfort and flexibility in the telephone arrangements. They provide alternate or additional locations for telephones if and when desired. By locating telephone jacks at these places, for use with a portable telephone, service could be provided in the rear bedroom when it was occupied by guests, or in the dining room, which would make possible a certain amount of privacy in conversation when the living room was occupied by others.

[15]

Illustration from American Telegraph and Telephone Company, *Planning for Home Telephone Conveniences: For the Use of Architects, Engineers, Builders, and Owners* (1928). Warshaw Collection of Business Americana, National Museum of American History, Smithsonian Institution.

and efficient service. In addition to the visible system, meaning the various speaking and listening points, the booklet included plans for the layout of the telephone conduits—or the hidden side of the network—because even if they are hidden, these conduits become essential to the functioning of the whole system and must therefore be precisely planned in advance. Moreover, while telephones are somewhat fixed and standardized objects, the layout of the conduits is the flexible and adaptable part of the system, and thus becomes strategic, conforming to specific needs and different spatial configurations.

One year later, AT&T published a second, less technical and more captivating version of its telephone equipment installation guide. This time, the booklet insisted on an explicit association with modernity and its needs, addressing the "modern American" and focusing on the "modern concept of comfortable living."[7] At the time, sound reproduction techniques represented a promise of

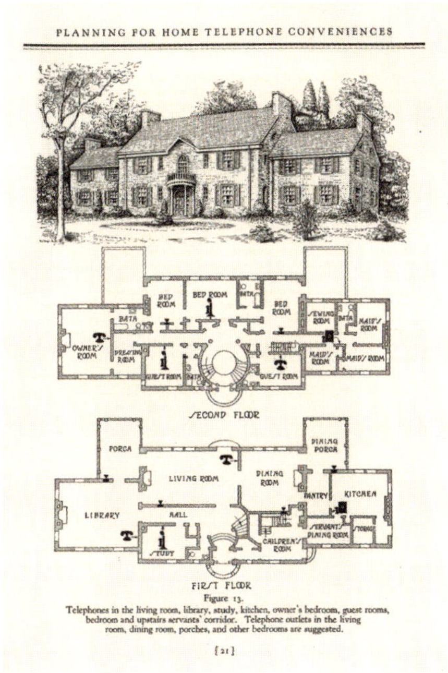

Illustration from American Telegraph and Telephone Company, *Planning for Home Telephone Conveniences: For the Use of Architects, Engineers, Builders, and Owners* (1928). Warshaw Collection of Business Americana, National Museum of American History, Smithsonian Institution.

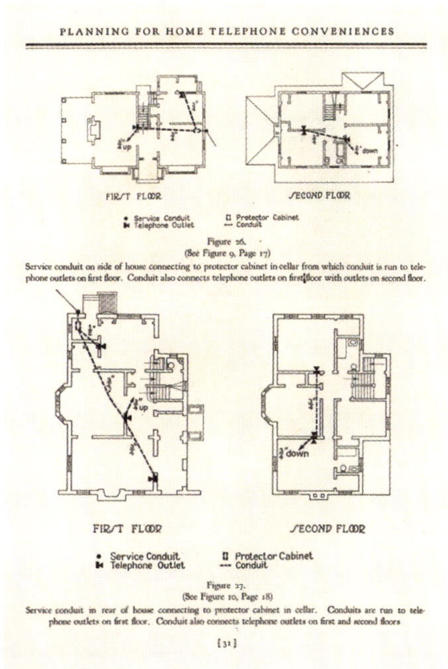

Illustration from American Telegraph and Telephone Company, *Planning for Home Telephone Conveniences: For the Use of Architects, Engineers, Builders, and Owners* (1928). Warshaw Collection of Business Americana, National Museum of American History, Smithsonian Institution.

science and rationality and introduced new relationships to everyday communication. In a word, they were considered "modern."[8] Moreover, devices such as telephones, radios, and the furniture that supported them embodied this relationship to modernity. They became "objects of desire," and their technical appearance a true emblem of progress.[9] But what is particularly striking in this case is that beyond the telephone itself, considered as an autonomous device, the real object of desire seems to have become the telephonic system as a whole, and how it fits into various architectural situations.

More than a simple technical manual, this second version of the booklet turns out to be a public relations document that introduces the aesthetic dimension of telecommunications equipment. "Such permanent built-in features have the double advantage of being both useful and decorative"[10]—and the authors once again insist on the adaptability of the service to different styles of

Illustration from American Telegraph and Telephone Company, *Modern Telephone Service for the Home* brochure (1929). Courtesy of AT&T Archives and History Center.

Illustration from American Telegraph and Telephone Company, *Modern Telephone Service for the Home* brochure (1929). Courtesy of AT&T Archives and History Center.

housing, from the modest one-story bungalow to the more articulate luxury dwellings, that are better served "by a complete intercommunicating service in addition to the ordinary facilities." The same series of plans for different types of houses is presented as an example: "A few typical floor plans will illustrate the way in which the modern American idea of comfort and convenience is being adapted to homes of many styles," the booklet states. Beauty must be combined with convenience in planning the proper setting for telephones and in the choice of the kind of furnishing to accommodate the telephone that fits the interior of the house. Tables, stands, and cabinets, it continues, "are available in an almost infinite diversity of styles and finishes. They may be had in choice woods of various colors and grains to harmonize with any scheme of home decoration." A remarkable axonometric view of a generic apartment building shows the arrangement of visible telephone devices and electrical outlets according to household needs. "With cities congested with traffic and apartments ever higher in the air,

Young people, too, have matters to discuss by telephone; matters which—to them, at any rate—are of more immediate importance than the most far-reaching financial transaction. Many of these matters are essentially private in their nature—particularly when there are still younger people in the household such as teasing small brothers or inquisitive little sisters. Somewhere in the home there may well be a telephone so located that it meets the personal needs of the grown-up yet still young members of the family. Older people, too, like to telephone, now and then, without being overheard.

Illustration from American Telegraph and Telephone Company, *Modern Telephone Service for the Home* brochure (1929). Courtesy of AT&T Archives and History Center.

Five pounds of . . . 2 cans of . . . a dozen . . . and have you any fresh . . . ?

The ordering of household supplies, expedited by telephone, makes up a large share of the important business of running the modern American home. The kitchen should have a telephone of its own, so located as to be instantly available when required, but out of the way when table space is needed for other work. Such installations also make it possible for the servants to answer incoming calls quickly and conveniently.

Illustration from American Telegraph and Telephone Company, *Modern Telephone Service for the Home* brochure (1929). Courtesy of AT&T Archives and History Center.

the modern standard of comfortable living is being reflected more and more by the number and location of telephones for the use of the family, servants, and guests." This mode of representation—the axonometric view—was often included in telecommunications documents of the time, most likely because it provided an effective means of displaying telecommunications installations and their integration into architecture, especially when considered as part of the planning process.

Another, more synthetic brochure on telephone installations presents an axonometric view of two empty and generic houses, one small or medium, and of a larger one as well, which are equipped with what the brochure calls "typical telephone conduit layouts."[11] Again, the authors' ultimate message is that the purchase or construction of a modern house cannot be done without providing for the installation of telephone conduits beforehand. These stylized versions of axonometric views, devoid of interior furnishings and other household equipment, further emphasize

Illustration from American Telegraph and Telephone Company, *Modern Telephone Service for the Home* brochure (1929). Courtesy of AT&T Archives and History Center.

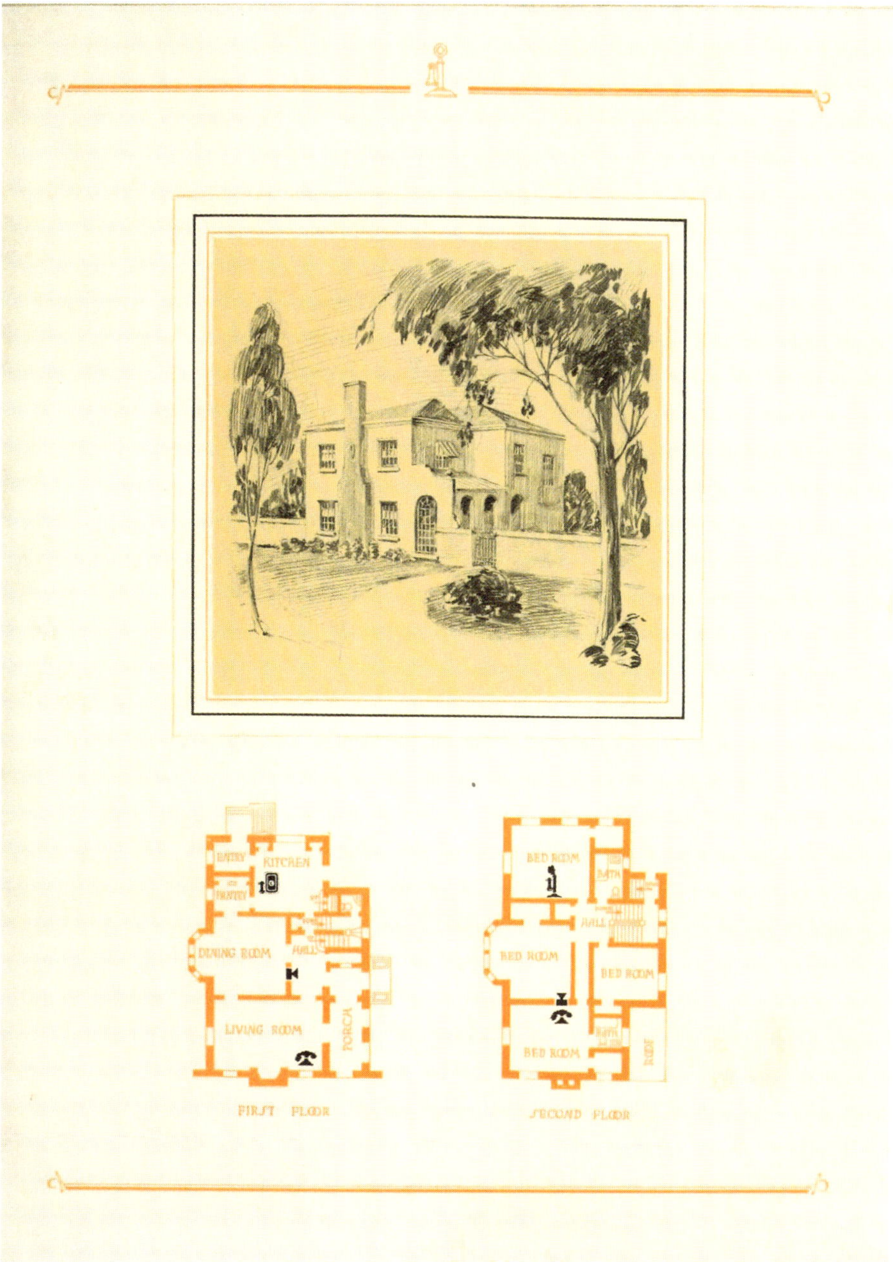

Illustration from American Telegraph and Telephone Company, *Modern Telephone Service for the Home* brochure (1929). Courtesy of AT&T Archives and History Center.

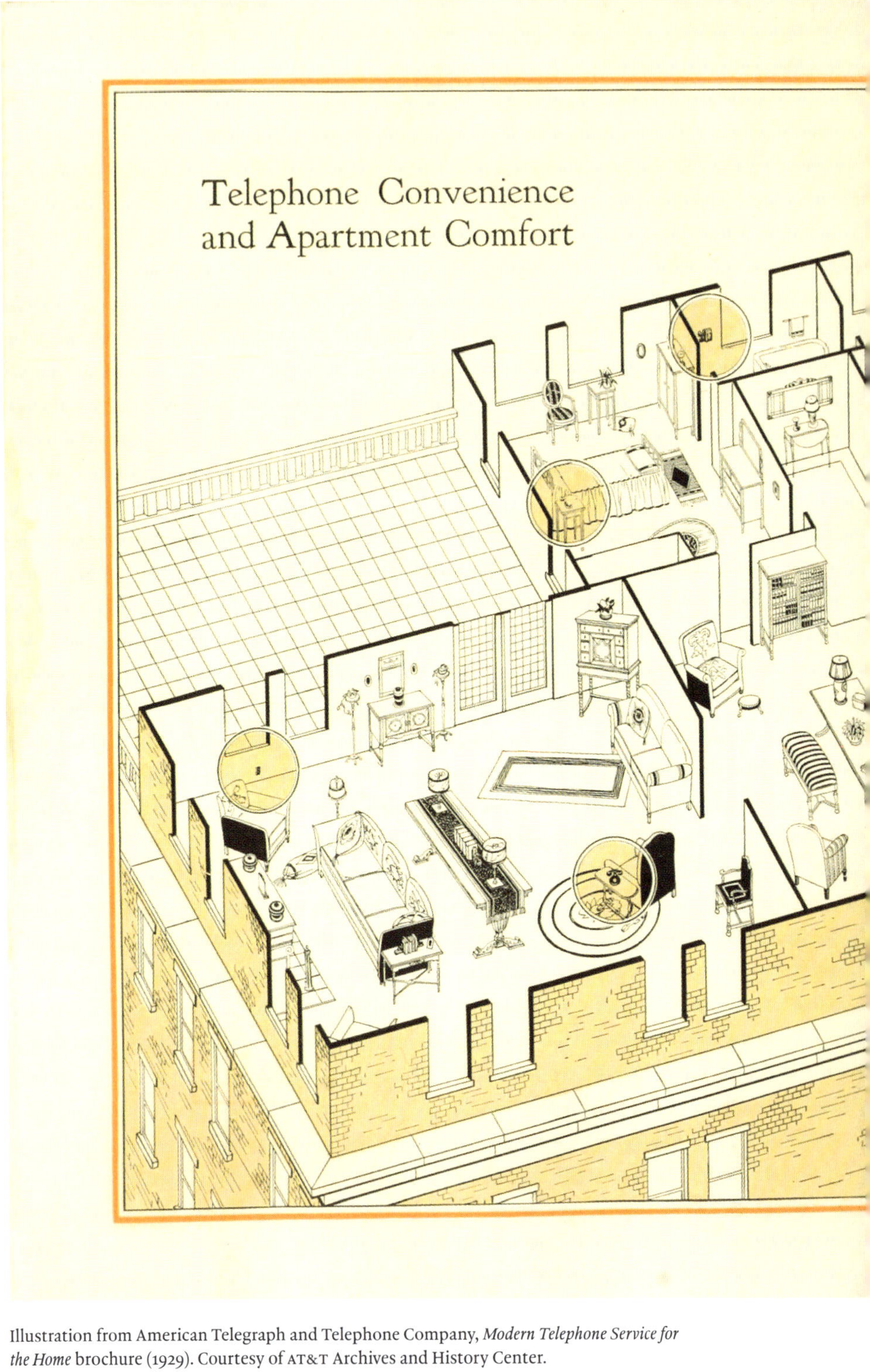

Illustration from American Telegraph and Telephone Company, *Modern Telephone Service for the Home* brochure (1929). Courtesy of AT&T Archives and History Center.

THE development of apartment house life, particularly in the large cities, has proved the convenience of telephone facilities that are adequate to the household's needs. With cities congested with traffic and apartments ever higher in the air, the modern standard of comfortable living is being reflected more and more by the number and location of telephones for the use of family, servants, and guests.

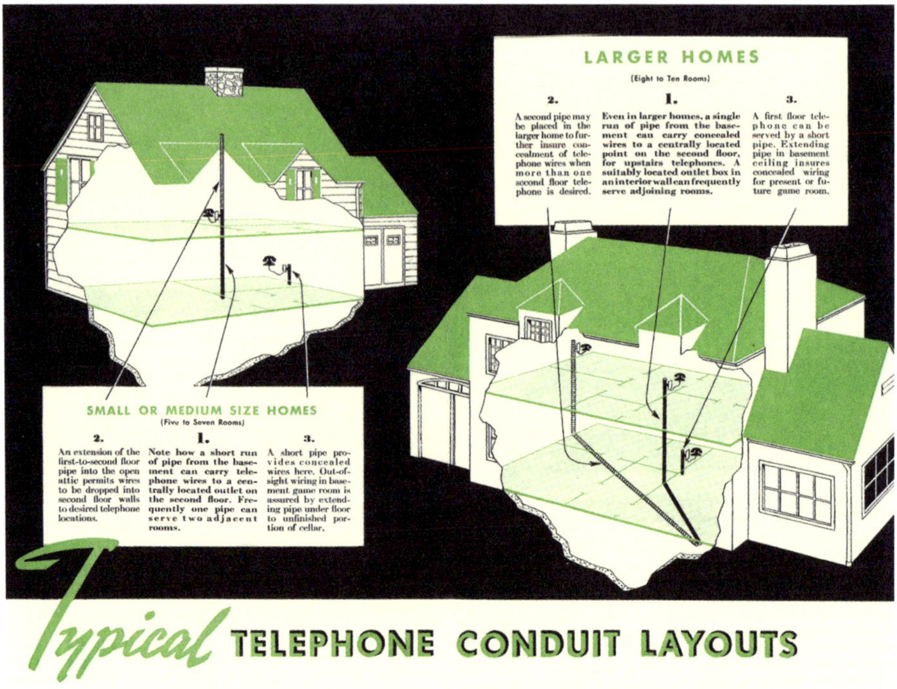

Advertising brochure for Bell System (New Jersey Bell Telephone Company), AT&T, n.d. Warshaw Collection of Business Americana, Archives Center, National Museum of American History, Smithsonian Institution.

the essential need to couple telephone installations with primary building elements such as walls, floors, and roof.

As Yve-Alain Bois and Bruno Reichlin have already observed—extending the Panofskyian tradition of interpreting symbolic form—axonometry is a mode of representation with an ancient tradition that became widespread in the modern era.[12] Reyner Banham observed how Gropius's famous isometric drawing of his own office at the Bauhaus followed a typically modern intention of emphasizing the design of heating and lighting systems as directly integrated into the volume and form of the room.[13] But he deplored any aesthetic use of axonometry that took insufficient account of the real conditions of use of the space, both environmental and human. Thus, a table lamp appeared in the photos of the realized project, as well as two telephones placed on a shelf in proximity to the radiator, to offset its technical deficiencies. According to Banham, Gropius used axonometry to show the technical functioning of the office, but without adequately responding to its actual use, especially in light of modern technology.

Axonometry is distinct from Renaissance pictorial perspective, as codified by Leon Battista Alberti and Piero della Francesca, the illusionist technique largely associated with a cultural model of its time.[14] Perspective focuses on the observer's perception of an object from some given spatial position, whereas axonometry preserves the actual dimensions of any given object to scale, regardless of its position in space. However, as soon as one leaves pictorial illusion and directly addresses the art of construction, one notices—already in the Renaissance—a crucial shift in the modes of representation. Around 1443, Alberti underscored the need to build models in *De re aedificatoria*, in order to go beyond "the appearance of perspective" and to verify the principles of construction based on reason and specific dimensions.[15]

A little further on, in the same chapter, he speaks of the importance of roofs and walls, but also of those elements that rise up from ground level or are dug down below it, such as conduits, canals, rainwater collection systems, sewers, and the like. The model provides an alternative to axonometry, another synthetic vision of the built object. It was already seen as a tool to show networks of circulation integral to the architecture—what we call infrastructure.

Axonometric projection was often used for technical drawings and deployed in the construction of military fortifications and machinery—whenever the detail, precision, and functioning of the objects concerned were a primary practical necessity.[16] It became a necessary tool, along with the model, for the classical architect of the Renaissance, as demonstrated by Leonardo da Vinci's "bird's eye view" drawings that provided a single representation of different parts of the building:[17] "this requirement implies a complete and definitive conception of the project before its realization."[18] No freedom of interpretation on the part of the mason would be conceivable from then on. But it was at the beginning of the twentieth century that the use of axonometry became widespread, extending well beyond the specialized and technical fields of construction, and came to embody a machinic conception of architecture.

The meaning of this diffusion is threefold. First, this mode of representation would become the expression of a new functionalist regime that induced a major cultural change—from a conception

Section perspective that shows the radio equipment and conduits installed in the interior of a house, n.d. George H. Clark Radioana Collection, Archives Center, National Museum of American History, Smithsonian Institution.

90131

of architecture as a static and formalized object to one that draws attention to its performative and mechanized processes. Second, unlike classical perspective, which limits the observer to a single point of view outside the building itself, the axonometric view allows the observer to enter the volume in order to have a complete image of it. And finally, from this interior point of view, the user becomes a kind of omnipresent figure, potentially dwelling (and communicating) in all the rooms of a house, from the bathroom to the living room, and so on. Axonometry does not offer a single, central point of view, but a set of equal views that suggest a temporal movement through the different interior spaces of a building, producing a "kinematic" perception of architecture.[19] Modern architects seeking to destroy the traditional supremacy of the facade have found axonometry to be an appropriate way to represent their buildings from multiple angles and to draw attention to the spatial and structural characteristics of their work. Because of its ability to represent the third dimension of architecture, axonometry also facilitates the display of different layers or phases of construction. Compared to a fixed, finite perspective, it tells us more about the planning process by revealing, for example, the presence of buried conduits in the walls of a structure and the various technical devices that occupy the architectural interior. Thus, this mode of representation makes explicit the interest of modern architects in technology by revealing the functioning of buildings and positing them as unified systems as a whole, spatially and technically.

Another document related to telecommunications, from the George Clark Radioana Collection (at the Smithsonian Institution), shows the conduits that feed a set of radio devices distributed in each room of the house, except for the stairs, by means of a perspective section. The electrical conduits, hidden in the walls and floors, allow different types of devices to transmit sound messages and music to a new type of domestic audience. A typical product of nineteenth-century culture, this type of analytical section presents "the interior as a façade," coupling the classical perspective that keeps the observer at a distance, outside, with the modernist need to represent technical services hidden in the architecture.[20] Through a form of graphic representation that highlights the tension between architecture and technique, as well as between architecture and construction, this document attests to a progressive interest in architectural representation on the part of

AT&T advertising, n.d. N. W. Ayer Advertising Agency Records, Archives Center, National Museum of American History, Smithsonian Institution.

telecommunications advertisements and also incisively illustrates a gradual transition to modernity.[21]

Thus, from the late 1920s onward, telecommunications company documents seem to borrow directly from architectural culture of the time. The widespread use of plans and axonometric drawings in advertising materials testifies to the desire to address professionals directly, as well as to make the technical language of architecture accessible to the general public. If technique—or the art of making things—is no longer something to be hidden, it becomes a sign of progress, an element to be emphasized and even enhanced. Yet, instead of drawing attention to the latest sound reproduction technology—in this case the telephone—this series of images emphasizes the way in which telephone conduits were planned directly in conjunction with architecture; it places particular emphasis on the articulated organization of these systems—and corresponding aspects of domestic life—shown in architectural plans and axonometric views. More generally, drawings function as mediators of architecture because they convey selected information about a real object. When a specific graphic language is selected, one chooses what particular content is to be maintained during the transition from the architecture itself (the object) to its mediator (the drawing).[22] Thus, on the one hand, the set of documents produced in the context of telecommunications

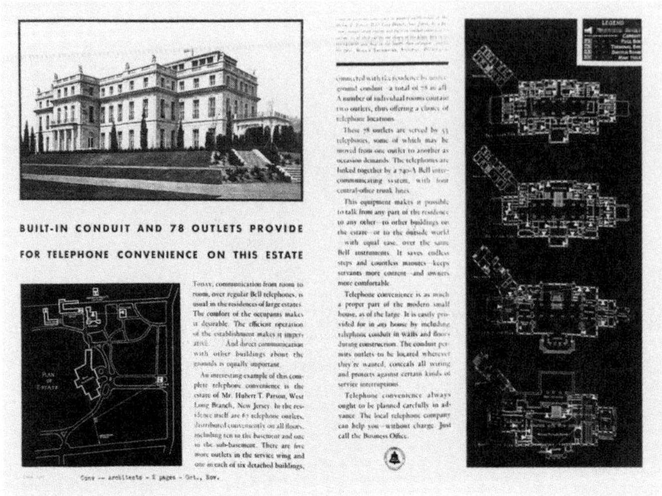

AT&T advertising, n.d. N. W. Ayer Advertising Agency Records, Archives Center, National Museum of American History, Smithsonian Institution.

AT&T advertising, n.d. N. W. Ayer Advertising Agency Records, Archives Center, National Museum of American History, Smithsonian Institution.

advertisements must be considered as an emblematic expression of the so-called machine age, which deliberately emphasizes the technicality of things. On the other hand, if architecture is a machine, as claimed by several authors, it is also a particular example. Devoid of cogs and mechanisms, architecture is a machine that is distinguished by its static condition, and becomes performative thanks to its infrastructure capable of producing movement in the observer and the user.[23]

Among the various interpretations of modernity at the time, two diametrically opposed positions emerged regarding the definition of the machine. The first focused on the sublimation of finite, geometric objects as plastic products with inherent symbolic values. This formal position was fully expressed, for example, in the 1934 *Machine Art* exhibition at the Museum of Modern Art (MoMA). In his introduction to the catalog, Philip Johnson quoted Plato's Philebus and stated that the "beauty of machine art is in part the abstract beauty of 'straight lines and circles' made into actual tangible 'surfaces and solids' by means of tools, 'lathes and rulers and squares.' . . . Machines are, visually speaking, a practical application of geometry."[24] This infatuation with geometry led Johnson, along with Alfred H. Barr Jr., to organize an exhibition on modern design that combined tools and machines, from

AT&T advertising, 1947. N. W. Ayer Advertising Agency Records, Archives Center, National Museum of American History, Smithsonian Institution.

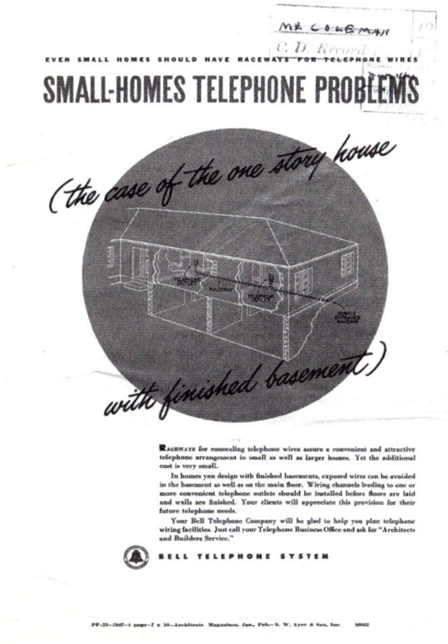

AT&T advertising, 1947. N. W. Ayer Advertising Agency Records, Archives Center, National Museum of American History, Smithsonian Institution.

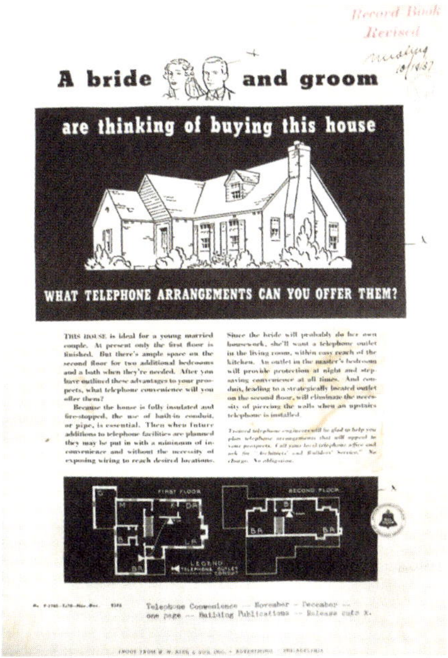

AT&T advertising, 1937. N. W. Ayer Advertising Agency Records, Archives Center, National Museum of American History, Smithsonian Institution.

AT&T advertising, 1938. N. W. Ayer Advertising Agency Records, Archives Center, National Museum of American History, Smithsonian Institution.

spoons to microscopes, without distinguishing between them and without really questioning how they worked. The second position focused on the operational behavior of the machine as the basis for asserting a more processual viewpoint. This position was well represented by another modern thinker. In the same year as the MoMA exhibition, Lewis Mumford published *Technics and Civilization*, in which he defined the machine as "the ensemble of knowledge and skills and arts implicated in the new technics."[25] The degree of independence of a machine from its operator—its automatic character—and the level of specialization of its functions were, according to Mumford, the criteria that determine the fundamental difference between a machine and a tool.

These two positions enter quite easily into direct dialogue with the corpus from the material culture of telecommunications. The first is linked to representations that rhetorically and almost plastically highlight the terminal sound device—the telephone, the radio. The second, which concentrates on the automation of functions and specialization of processes, refers to representations that emphasize the underlying technical systems as well as to implied spatiotemporal relations and practices—as does the specific corpus drawn from advertising documents of the telecommunications companies described earlier. In this sense, the representations of conduits in architectural interiors offered a machinic reading of telecommunication and, therefore, of architecture. More than sound reproduction devices as isolated objects, sound conduit systems ultimately reveal the processes and workings of a new era of communication that, starting from the domestic scale, connects to the potentially limitless scale of the territory.

8 *Public:* Au théâtre chez soi

"It's all very curious. One listens to the performance of the *Opéra*. One changes ear-covers and hears the *Théâtre français*, Coquelin, etc. One changes again and hears the *Opéra-Comique*. The children were charmed and so was I."[1] In one of his notes dated November 11, 1881, Victor Hugo described with some amusement the discovery of a new electric device, the *théâtrophone*, enabling people to listen in real time to performances from the great Parisian theaters, while facing an upholstered wall.

In the context of Parisian society toward the end of the nineteenth and the beginning of the twentieth century, the théâtrophone was a particularly interesting invention of the modern era in the field of sound recording and reproduction.[2] Between infrastructure at the scale of a city, and an individualized listening device, and as a remarkable social invention for the audiophiles of the time as well, the théâtrophone called the theatrical experience itself into question, for the audience was physically removed from the place of performance.

If this dissociation of the senses—one hears but is physically elsewhere—is an artifact of the process of technical specialization of the modern era, the théâtrophone nevertheless retained an important societal value: one could listen in real time to what was taking place in the great halls of the city and be virtually present at the very moment of occurrence, often sharing the moment

Théâtrophone Company advertising, 1894. Paul Charbon and Pierre Lajarrige, "Le téléphone ce jeune centenaire: le Théâtrophone," *Revue française des télécommunications* (October 21, 1976): 52. Bibliothèque historique des Postes et des Télécommunications. Orange/DNAP.

of listening in other venues. The elite class of theater audiences was essentially extended to a community of curious and cultured technophiles. Most often forgotten by the history of telecommunications, this invention nevertheless developed over some fifty years to entertain a new public in its search for new experiences.

The théâtrophone introduced the practice of acoustic delocalization, which has been characterized, among other things, as the "*théâtre chez soi*." This concept corresponds to its second phase of development, when the device literally entered the domestic space, whereas the théâtrophone was initially designed for shared public spaces other than the theater. The expression "theater at home" or "home theater" has indeed been associated with several modern media (telephone, phonograph, radio), or with musical scores for amateurs that allowed collective listening inside a domestic space even earlier.[3] However, in the case of the théâtrophone, the expression was no longer an allegorical image, but a direct attempt to transmit the sound of the theater by electrical means inside one's own home.

In other words, the théâtrophone was an invention of the modern bourgeoisie. In the words of Matei Calinescu, the notion of "bourgeois modernity" is associated with the doctrine of progress, with confidence in the beneficial possibilities of science and technology, with an orientation toward pragmatism, and with the question of time in particular, a measurable time that can be bought and sold, a time that becomes a commodity.[4] The théâtrophone embodies this bourgeois conception of progress, through a reification of the time of the spectacle and of culture. It is an invention that is inscribed "at the apogee of capitalism," to use one of Walter Benjamin's expressions. The modern man seeks "refuge in the mass of the big city," to isolate himself at the very moment when he is in contact with it.[5] The théâtrophone, as it was originally conceived, is one answer to this quest for individualization inside the mass-produced by the mechanization of the culture, and as described by Benjamin.

In 1879, Clément Ader, the inventor of the théâtrophone, participated in the installation of the first private telephone network in Paris. During the International Exhibition of Electricity of 1881, he presented for the first time "telephone auditions" that allowed one to listen to live performances coming from the *Opéra* and the *Comédie française*.[6] In two different rooms of the Palace of Industry, some forty listening posts were to be found, a sort of ancestor of the telephone booth, with double-earphone devices manufactured by the Ateliers Breguet. These isolated each listener in their own sound bubble, facing an absorbent wall. The rooms were all covered with hangings and carpets. They included two doors covered with thick fabric, a dim light to minimize visual distraction, and a table in the center, to accommodate a supervisor who cut the sound transmission circuit every five minutes. A new group could thus enter and experience an immersion in the sound space of the great Parisian theaters.

Invented in 1876, the théâtrophone was one of the first applications of the telephone. But its system was based on the dissociation of its two principal functions: the receiver on one side, and the transmitter on the other. The network was composed of a series of microphonic sensors located on the theater stage that transmitted the sound signal through underground wires to the listening posts of the Palace of Industry (and also to the similarly

Telephone listening room at the Palais de l'Industrie (Palace of Industry) at the International Exhibition of Electricity, engraving, Paris, 1881. "L'Exposition d'électricité," *La Nature*, no. 434 (September 24, 1881): 17. Cnum—Conservatoire numérique des Arts et Métiers.

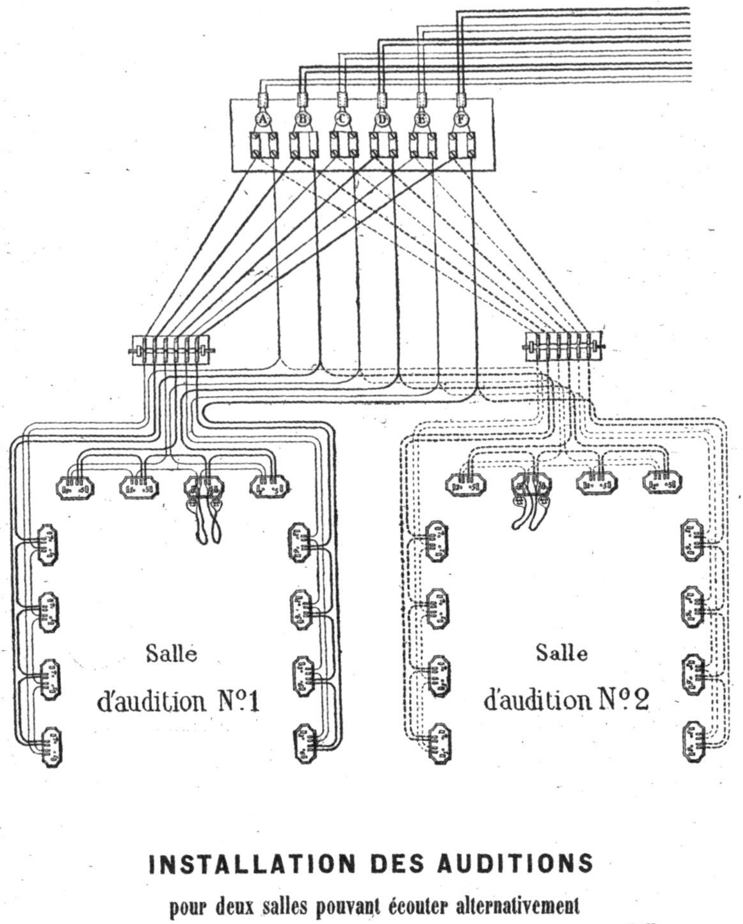

Layout of theatrical auditions at the International Exhibition of Electricity, Paris, 1881. Julien Turgan, *Les grandes usines, études industrielles en France et à l'étranger* (Paris: Calmann Lévy, 1882), 72. Cnum—Conservatoire numérique des Arts et Métiers.

equipped Élysée Palace, around one mile away, that allowed President Jules Grévy to inaugurate the system). The théâtrophone is thus part of a family of modern inventions (telegraph, telephone, radio) that do not rely solely on the design of a device, but require setting up a network, an infrastructure. The system was a response to a possibility imagined at the dawn of the development of the telephone—an application dedicated to entertainment and culture that would later be taken up by other technologies, such as the radio and the phonograph, but which was part of Parisian bourgeois society for almost half a century.

According to Jacques Attali and Yves Stourdzé, the adoption of the telephone in the French political context amounted to a kind of ideological revolution, through its capacity to introduce a two-way dialogue into the domain of communication, a domain that had traditionally been administered as a one-way monologue, controlled by the state. In this context, the *théâtrophone* represented a much more controllable and acceptable application of the potentially open network of the telephone, still intended for an elite audience.[7] The nationalization of telephone service in France took place in 1889, one hundred years after the French Revolution, in a society that had been built on principles of order and hierarchy. The initial delay in the development of French telecommunications was due, Attali and Stourdzé stated, to resistance regarding a network that escaped the total control of the State over its contents and operations. The French invention of the Chappe telegraph, or semaphore, in 1794 marked an important milestone in the prehistory of telecommunications and was intended by its inventor Claude Chappe to transmit one-way messages from the Government to the Republic.[8] The infrastructure of this first telegraph network was conceived as a system of quasi-military architecture: towers supported by walls, doors and armed men who protected the actual devices, which consisted of a mast and arms for the visual transmission of messages. When the first lines for the electric telegraph were introduced in France, in 1846, a collective apprehension was expressed regarding an infrastructure left without guards and consisting of wires and posts easily subject to sabotage, breaches, disconnection, and external manipulation.

Returning to the théâtrophone and its relatively limited history of operation, one reads how this largely unidirectional network for leisure time, (from the theater to the audience) seemed a more obvious and "reassuring" form of use than the free and interpersonal communication of the telephone. The théâtrophone appears in several narrative accounts, with technophilic enthusiasm being one of its main sources of attraction for the writers of the time.

In his novel *Looking Backward*, published in 1888, Edward Bellamy described a very similar invention, a musical and liturgical program, timed according to the hours of the day, which was broadcast by the telephone network in hotels, restaurants, and private apartments. "Now, as to hearing a sermon today, if you wish to do so, you can either go to a church to hear it or stay at home," says Julian West, the novel's protagonist who finds himself in a world transformed after one hundred and thirteen years of sleep. "There are some who still prefer to hear sermons in church, but most of our preaching, like our musical performances, is not in public, but delivered in acoustically prepared chambers, connected by wire with subscribers' houses. If you prefer to go to a church I shall be glad to accompany you, but I really don't believe you are likely to hear anywhere a better discourse than you will at home. I see by the paper that Mr. Barton is to preach this morning, and he preaches only by telephone, and to audiences often reaching 150,000."[9]

The imaginary underlying these pages of the utopian novel emphasized the increased possibilities of listening at home, in a place acoustically adapted, as well as being able to reach a potential and invisible audience on a mass scale, and not simply the limited public of the traditional theatrical experience. Beyond fictional accounts, in the United States (as well as in other countries

Humorous vignette on the use of the radio "chez soi," 1946. George H. Clark Radioana Collection, Archives Center, National Museum of American History, Smithsonian Institution.

such as Belgium, Germany, Austria England, Hungary, and Italy), experiments similar to the théâtrophone also existed and were supported by the Long-Distance Company.[10]

Thrilling for some, as evidenced by this type of narrative, the experience of listening at a distance was more disappointing for others, for whom the theater still represented a social occasion of public display. One went there to attend a performance and also to be seen by a select layer of society, to share these moments of public leisure. A humorous vignette that refers in this case to radio broadcasting shows an elegantly dressed couple, ready for a theater session, sitting on an upholstered sofa in a domestic interior space. In the foreground is a radio set on a carefully decorated coffee table. The man seems to be listening attentively to the program, with his eyes closed, while enjoying hors d'oeuvres. The wife, on the other hand, seems bitterly disappointed by the acousmatic performance her husband invited her to.[11]

The modern technologies of sound reproduction made it possible to replace an original source, like the musical performer, with a technical object that itself would become the basis for a new sonority. This new tension between the cause and the sound effect was summarized by Brian Kane as acousmatic listening, and it is in fact central to the théâtrophone device, which combines the powers of transmission of the new electric technologies with the more traditional societal requirements that still require live listening of the concert.[12]

In the illustration, one can imagine that it is not only the fragmentation and qualitative impoverishment of the perceptual experience of the performance that is to be deplored, but also the loss of the ability to manifest one's own physical presence as an audience—in a ritual dialogue between hall and stage—that is brutally amputated by the suppression of the real theater space.

It is also evident, in the imaginary associated with the théâtrophone, and in the advertising images and various illustrations, that the public is constituted to a great extent by women. The safe and glamorous aspects of the device were emphasized by commercial publicity campaigns, at a time when electricity and its introduction into the domestic habitat were still seen as potential sources of danger.[13] The feminine incarnation of the use of electricity made it possible to gain the confidence of a segment of the public that was considered more vulnerable, and to use the expression of a woman's body and clothing to attract attention.

Jules Chéret, *Théâtrophone*, lithographic poster, 1890. Artefact / Alamy Stock Photo.

Woman with a telephone, 1893. Warshaw Collection of Business Americana, Archives Center, National Museum of American History, Smithsonian Institution.

Audition of the théâtrophone in the salon of a grand hotel in Paris, 1892. Georges Mareschal, "Le Théâtrophone," *La Nature*, no. 995 (June 25, 1892): 56. Cnum—Conservatoire numérique des Arts et Métiers.

These images are emphasized in order to maintain the social role of the audience, which, as in the theater, wants to be seen by a community of aficionados.

Moreover, a specific choice of genres regarding the use of the telephone would soon emerge. At the outset, the network developed in the "male" business milieu, but the system quickly extended to the domestic environment, with an affirmation of new "feminine" practices of remote conversation, promulgated by company advertisements.[14] The introduction of party lines in the 1880s was explicitly aimed at a female audience of the bourgeoisie or middle class, and it opened up the conversation to a group of subscribers. An emerging "telephone culture" promoted new forms of sociability, especially in rural areas, that was less intimate than listening to the telephone in pairs, and that was often associated with the exchange of gossip, with "living room conversation" and with passive listening as well.[15]

Returning to the théâtrophone, it was the introduction of stereophony that most distinctly marked its invention from a point

AT&T advertising, n.d. N. W. Ayer Advertising Agency Records Archives Center, National Museum of American History, Smithsonian Institution.

AT&T advertising, 1933. N. W. Ayer Advertising Agency Records Archives Center, National Museum of American History, Smithsonian Institution.

AT&T advertising, 1932. N. W. Ayer Advertising Agency Records Archives Center, National Museum of American History, Smithsonian Institution.

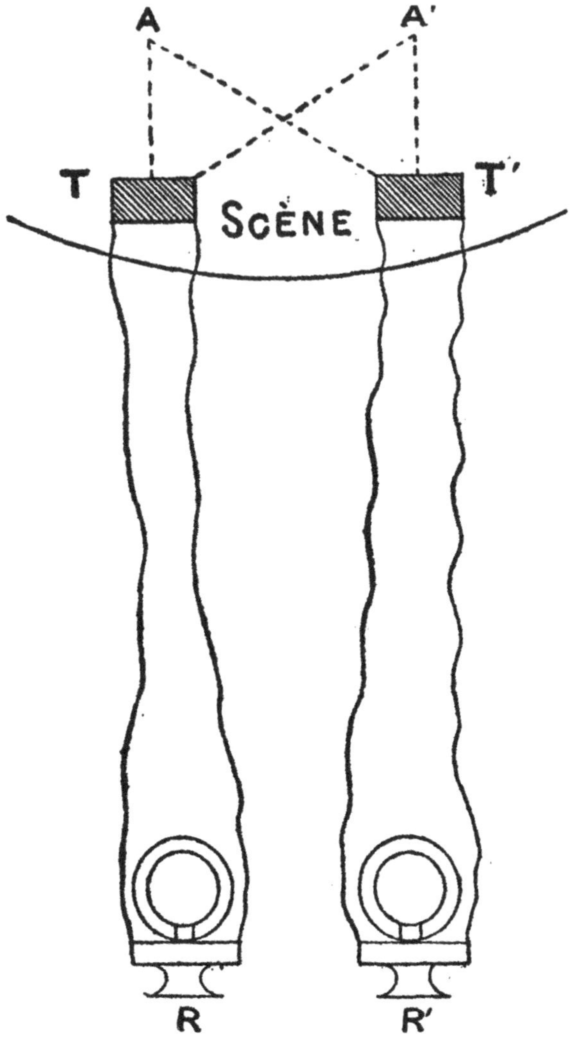

Operating diagram, stereophonic listening to the théâtrophone: "You were at the Palais de l'Industrie and yet you were attending the Opera," 1882. Julien Turgan, *Les grandes usines, études industrielles en France et à l'étranger* (Paris: Calmann Lévy, 1882), 68. Cnum—Conservatoire numérique des Arts et Métiers..

of view of listening practices. In an effort to reproduce the "realistic" space of the stage, Ader's device allowed for binaural listening: the two earpieces were connected to two microphones located on the ramp, at the front of the stage, to the right and left of the prompter's box.[16]

As with nineteenth-century optical devices, such as the stereoscope, which, by relying on the principle of binocular vision, reproduces "realistic" spatial effects while in truth passing through a form of abstraction and standardized reconstitution of optical experience, the théâtrophone also proposed a synthetic form of restitution of three-dimensional space.[17] The movement of the scene was thus perceived by the movement of the sound from one earpiece to the other. Moreover, as in the real space of the theater, where our unique visual and auditory perception of the performance depends on our location, the listener of the théâtrophone also had a unique individual perception, because at the origin of the system, each microphone had a corresponding listener (an equivalent series of microphones was thus placed in a row on the ramp, the number of listeners being therefore limited). Even while being elsewhere, one could preserve several physical conditions related to the theatrical experience, like the movement of the interpreters on the stage—a form of spatial restitution of the sound with the conservation of the noises and the ambient effects of the place—as well as the specificity of a distinct point of listening for each spectator: "the old subscribers to the théâtrophone still ask that they be placed on a good row, 'on the side of the violins,'" said M. Testavin, the last director of the théâtrophone company, who in 1930 reported on the evolution of this system.[18]

The dissociation of the senses and the fragmentation of experience produced by this instrument once again conformed to the process described by philosopher Walter Benjamin as an obvious consequence of the modernization of the work of art: "Thus technique has subjected the human sensorium to a training complex."[19] If Benjamin identified Baudelaire as the poet of the crisis of the modern city, obsessed with the image of the shock provoked by the crowd of the big city, and in search of isolation by poetry, he also saw in Baudelaire's situation evidence of the irremediable impoverishment of the experience of the work of art, "the price that the modern man has to pay for his sensation: the collapse of the aura in the lived experience of the shock."[20] And it is also a shock, or a consistent temporal contraction, that occurs in the modern listener of the théâtrophone, who is immersed only for five minutes in this sound space transmitted by electric wires. The theatrical spectacle is no longer lived in full, neither on the level of the body, of the senses, or the space, nor on the level of time, with

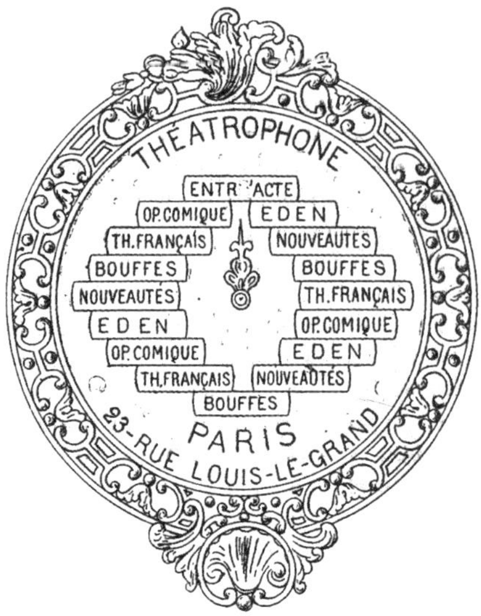

Théâtrophone dial, 1894. Julien Lefèvre, *L'électricité au théâtre* (Paris: A. Grelot, 1894), 334. Bibliothèque nationale de France.

A model of théâtrophone, 1893. Louis Montillot, *Le téléphone pratique* (Paris : A. Grelot, 1893), 465. Bibliothèque nationale de France.

its ceremonial and its social codes (arrival to the theater, waiting, intermission, lowering of the curtain, applause, curtain calls, exit from the hall). It is a partial restitution that allows a glimpse of the spectacle, a sample that anticipates current forms of compulsive listening; one consumes short extracts of pieces, one changes channel, frequency, or file, to seek a maximum concentrate of pleasure, what commonly would be called "channel surfing."

The subsequent model of the device, exhibited this time on the occasion of the Universal Exhibition of 1889 by Bélisaire Marinovitch and Geza Szarvady, was portable and accepted tokens (fifty centimes for five minutes of listening, one franc for ten minutes), in the manner of a jukebox. With the creation of the Théâtrophone Company (*Compagnie du Théâtrophone*), its central office located at 23 rue Louis Le Grand, a fully functioning network was established with several receiving devices in cafés, restaurants, clubs, and hotels located on the grand boulevards, allowing one to listen to several shows taking place in different theaters in the city. Starting in 1891, the company offered a subscription service that allowed the sound of Parisian theaters to be brought into the private space of homes, either through machines supplied by the company or through a telephone network that was already installed, marking

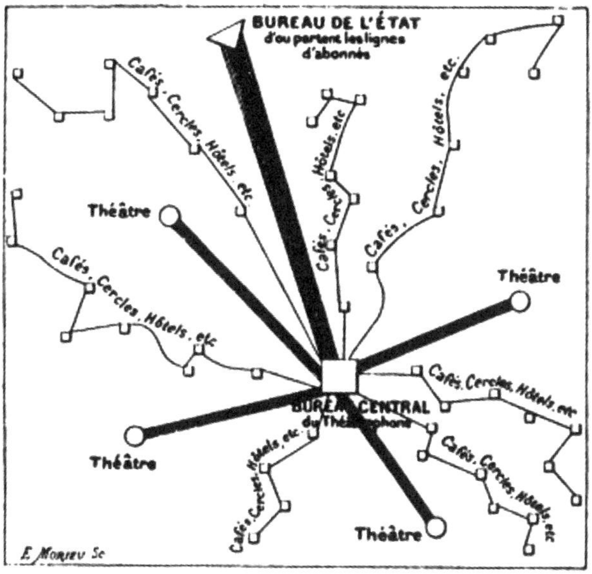

Diagram of the diffusion of the network of the théâtrophone in Paris, 1892. Georges Mareschal, "Le théâtrophone," *La Nature*, no. 995 (June 1892): 56. Bibliothèque nationale de France.

a return to monophonic listening. From the 1920s onward, Sunday sermons from the cathedral of Notre-Dame supplemented the service's programming. According to several sources, the annual subscriptions were aimed at a rather well-to-do public and made it possible, for an additional price, to free oneself from listening time limits, as well as from the number of halls one could access with the device. The choice of performances was made by a *théâtrophonist*, a kind of musical "operator," who selected the different communication circuits subscribers were able to listen to.

Mediocre sound reproduction quality was the subject of a critical response from a very famous théâtrophone subscriber: Marcel Proust. In 1911, he saw an advertisement published in the *Tout Paris* promoting "Le Théâtre chez soi," stating: "For auditions of: opera-opéra-Comique-Variétés-New releases-Comédie Française-Colonne Concerts-Châtelet-Scala, contact Théâtrophone 23, rue Louis-le-Grand Tél. 101-03 Subscription price for three people to attend daily auditions: 60 F per month. Trial audition on request." Proust subscribed to the service and became a lukewarm promoter of it. On the one hand, he was delighted to listen to Wagner's operas, of which he was a connoisseur, and to be able to "make up

Central office of the Théâtrophone Company, 23 rue Louis Le Grand, Paris, 1892. Georges Mareschal, "Le théâtrophone," *La Nature*, no. 995 (June 1892): 57. Bibliothèque nationale de France.

for the [théâtrophone's] acoustical deficiencies" thanks to his personal recollections of the works.[21] On the other hand, he deplored the poor acoustic reproduction that limited his use. Not without humor, he pointed out a confusion of sound levels between the stage and the hall that interfered with listening to the "musical heresies" of the avant-garde composer of his time, Claude Debussy. He noted with some enthusiasm how "at one point I found the rumble pleasant but yet a little amorphous, until I realized that it was intermission."[22]

In the 1920s, technical advances would actually impoverish the sound quality of the original théâtrophone experience. The multitude of microphones on stage were reduced to a single sensor capable of broadcasting on several networks. The individual quality of each recording device was replaced by a centralized system. On the reception side, a form of fatigue using the individual headphone became increasingly common, especially over long periods of time. Broadcasting passed increasingly to loudspeakers, especially the T30 model, the same one used for radio in the

1920s and 1930s. In other words, the "realistic" quality of stereo listening was replaced by monophonic listening, which nevertheless allowed increased access to the system, which extended to three hundred subscribers in 1930.[23] However, competition with other networks and technologies of the time was one of the main causes of the decline of the théâtrophone.

> The mentality of the théâtrophone subscriber differs essentially from the wireless transmission enthusiast: the latter finds great pleasure in the many small manipulations and adjustments required for the operation of a wireless receiver and is very interested in the new technology, and is often more interested in them than in the musical quality of the reception itself. The subscriber to the théâtrophone on the contrary, only considers the musical side: he much prefers to have nothing to adjust and to be relieved of all maintenance worries. Our choice is therefore made; the amplifiers will be placed at the central station and handled by competent hands.[24]

At a time when wireless technology, phonography, and the théâtrophone were still being developed, Testavin was still betting on the latter and assuming that the number of technical intermediaries (electromagnetic waves, amplifier, stylus, groove, etc.) between microphone and loudspeaker would have an impact on the future development of the systems, stating that "the fewer the number of intermediaries, the more faithful the reproduction."[25] However, history would reveal just the opposite: the "more direct" system (microphone—wire—loudspeaker) of the théâtrophone would disappear in favor of more elaborate ones, such as the radio and the phonograph, which would remain a part of the listening practices of the twentieth century much longer.

In other words, the bourgeois project of "realistic rendition," in which the listener was *as if* in the theater, would be overtaken by more interactive relationships of sound offered by technologies such as the radio and the phonograph.[26] The free manipulation of frequencies and sound storage would be a major factor in the development of listening practices. This was an open area of exploration that would eventually captivate the modern audience and take the place of the passive listening of the théâtrophone, even if the sound quality of latter had been superior at the time.

Illustration by Norman Rockwell, *Saturday Evening Post* (April 29, 1944). George H. Clark Radioana Collection, Archives Center, National Museum of American History, Smithsonian Institution.

A letter dated April 6, 1936, outlines the official request of the director of the Paris Opéra, Jacques Rouché, to withdraw théâtrophone equipment from the establishment.[27] The broadcasting of musical programs by telephone lines would later be the object of other types of use with, for example, the creation of Muzak, a company for the broadcasting of ambient music dedicated to social engineering applications in public spaces (shopping centers, parking lots, elevators) and workplaces (factories and offices).[28]

This system, patented in 1922 by General George Owen Squier, and marketed under the Muzak label in 1934, originally used the telephone network. For a dollar and a half, one could listen to the latest hits performed by the company's orchestra. The principle of passive listening was in this case put forward by this service, which beyond the technical invention that would quickly evolve toward the radio technology, aimed at filtering the musical content in such a way as to transform it into a background sound. Muzak largely anticipated a modification of modern listening practices by assimilating forms of audience inattention, while the théâtrophone remained more directly linked to a revered form of theatrical performance, while suppressing several of its codes.

In this sense, the théâtrophone was a transitional invention between evolving cultures and social practices, between the rite of the theater primarily addressed to the *haute bourgeoisie* or even to the aristocracy, and the modern technophile pleasure that allows anyone to participate in the diffusion and partial restitution of this experience. The latter practice responds better to the expectations of the middle class, particularly that of a growing petty bourgeoisie made up of music lovers and technical tinkerers.

Instead of undergoing a progressive technical evolution that continues today, as in the case of the telephone, the history of the théâtrophone was marked by the eclipse of its system. After half a century of use and repeated attempts to popularize the service, the théâtrophone was completely replaced by similar but better positioned technologies in the era of culture massification.

9 *Booth: The Public Post*

The telephone booth brings together several facets of the modern telecommunications system: small, standardized architecture, wired infrastructure, private means of communication, and public urban service. In the grid of telecommunications infrastructures, the phone booth represents a specific space—a bubble, an individual sphere, distributed throughout the territory. It differs according to the country, the telephone company, the climatic conditions, and the specific context of its installation (interior or exterior), but today the telephone booth is an object in the process of extinction or adaptation, and it is rapidly being replaced by the spread of mobile telephony. In some countries, the booth's architecture becomes an advertising medium, taking advantage of public legislation that still gives it an essential role and function. In other cases, booths host connected digital platforms; they are used to house surveillance systems,[1] or even more autonomous and occasionally creative purposes such as small art galleries, aquariums, library booths, spontaneous sanctuaries, money dispensers, or defibrillators.[2]

To understand the genealogy of the public pay telephone, it is necessary to distinguish the technical object from its architecture: a kiosk, an aedicule, a shell, or just an open structure supporting a roof. While furniture was built to house private telephones from the very beginning of the invention, some public telephone systems were also tested without necessarily being housed in protective constructions.

It is again at the 1881 International Exhibition of Electricity in Paris that the first public demonstration of a telephone booth seemingly took place. The private General Telephone Company (*Société Générale des téléphones*) installed thirty devices in upholstered oak cabins, which according to curious visitors looked like "confession booths."[3] The telephone was placed horizontally on a desk, making the manipulation of it quite uncomfortable until Clément Ader, the inventor of the théâtrophone and a pioneer of aviation, developed a vertical wall phone. The same year, another early German prototype (a "telephone kiosk") was installed at Französische Strasse in Berlin and operated by purchasing tickets in advance for a five-minute communication.[4] Introduced in 1899, the coin-operated telephones developed in Germany were produced in a variety of materials and shapes. In 1913 the kiosk became standardized with a showy hipped roof, large glass windows, and the sign "Automatic phone."[5]

In the United States, the first patent for a pay phone device was filed in 1888 and registered on August 13, 1889, to William Gray, an American precision machine polisher, hobbyist, and sports enthusiast, who had previously invented an inflatable chest protector for baseball catchers.[6] According to a story circulated by the company Gray's would eventually create, the idea for a public telephone came to him when his wife was ill and he could not find an available telephone to call a doctor. With the local firm of Pratt & Whitney, Gray's employer for several years, he made a first model that was installed on the ground floor of a bank at Main Street and Central Row in Hartford, Connecticut, by the Southern New England Telephone Company. As there was no cabin containing the device, the telephone remained exposed. Two years later, the Gray Telephone Pay Station Company was born. New models followed in succession. An initial system of postpayment was considered too constraining and was replaced by one that depended on advance payment on a multi-part device; Gray would end up filing for approximately twenty patents.

Several years earlier, a patent for a rectangular (1.2 × 1.5 m) or a circular telephone booth had already been filed by Richard Busteed of Brooklyn, New York, on August 7, 1883.[7] It was to be made of wood or metal, placed on wheels, equipped with a shelf and material for writing and hanging messages on the inside, with vertical ventilation and an external signal to indicate the presence of the user. This small construction on wheels was intended to create a place of individual accommodation. No messages were to leave the room; no nuisances were to enter from the outside, and it could be moved about within a larger public space.

The first outdoor telephone booth would be installed in the United States in 1905 by the Bell System, on a street in Cincinnati.[8] The telephone booth explicitly embodies the possibility of isolation within the very large system offered by modern telecommunications. Whether it is completely enclosed by doors and glass, or is simply acoustically isolated under a shelter, the telephone booth tends to create a true bubble of acoustic privacy, even if the user can be seen through the transparent glass (a variation that did not arrive in the United States until the 1950s), or a part of his or her body can project from the envelope around the listening device. In this case, the visibility of the user becomes a signal that the device is in use and that the booth is already occupied.

In Europe, the installation of phone booths depended on various companies and legislations. Starting in 1885, oak booths upholstered in moleskin and equipped with adjustable velvet armrests were installed in the French network of post offices and central telephone offices. They required the mediation of a PTT (*Postes, Télégraphes et Téléphones*) employee to make calls, collect payments, and to supervise the general operation.[9] Human mediation of the apparatus, especially by women, characterized the French system during the first decades of its development, generating a hybrid form of machine directly connected to individual control of the service staff.[10] As with the theâtrophone, the figure of the communications "manager" differed from the more common switchboard operator who was responsible for connecting the wires on the board and was in remote voice contact with the audience. This role persisted longer in France than elsewhere and indicated a certain inability to think of the telephone as an autonomous mass communication tool from the outset.[11]

With the previous technology, the telegraph, men had been the only ones to manually take care of the connections and interpret

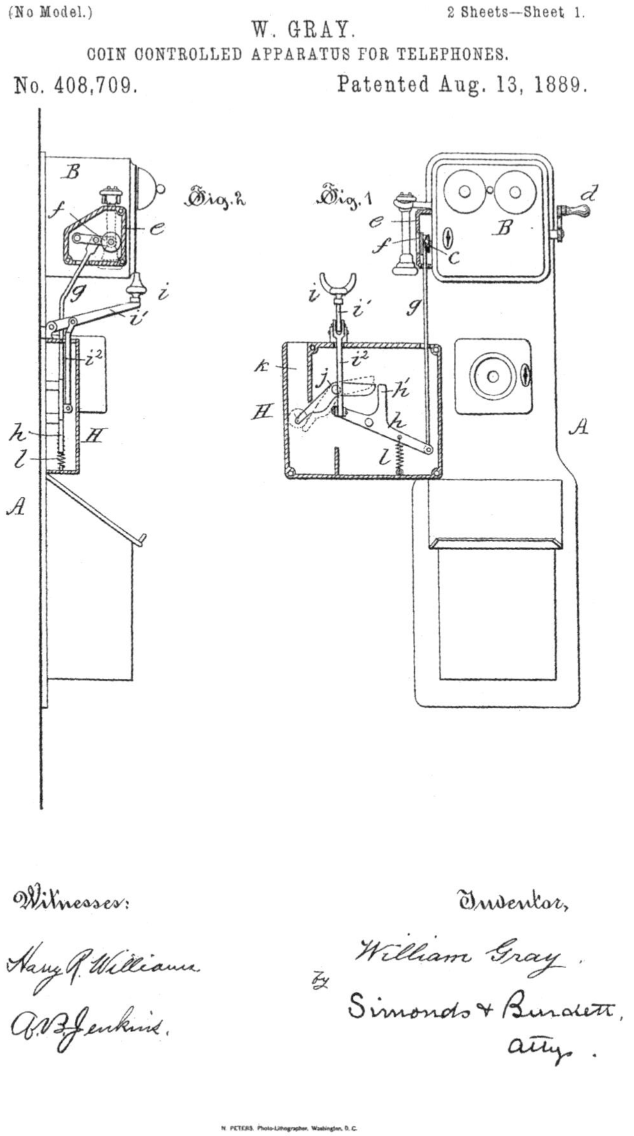

William Gray, Coin-Controlled Apparatus for Telephones, no. 408,709, patented August 13, 1889.

the signals of the telegraph. But with the advent of the telephone, the figure that acted as a mediator was completely feminized. After the first female switchboard operator, Miss Emma Nutt, was hired in 1878 by the Boston Telephone Despatch (sic) Company

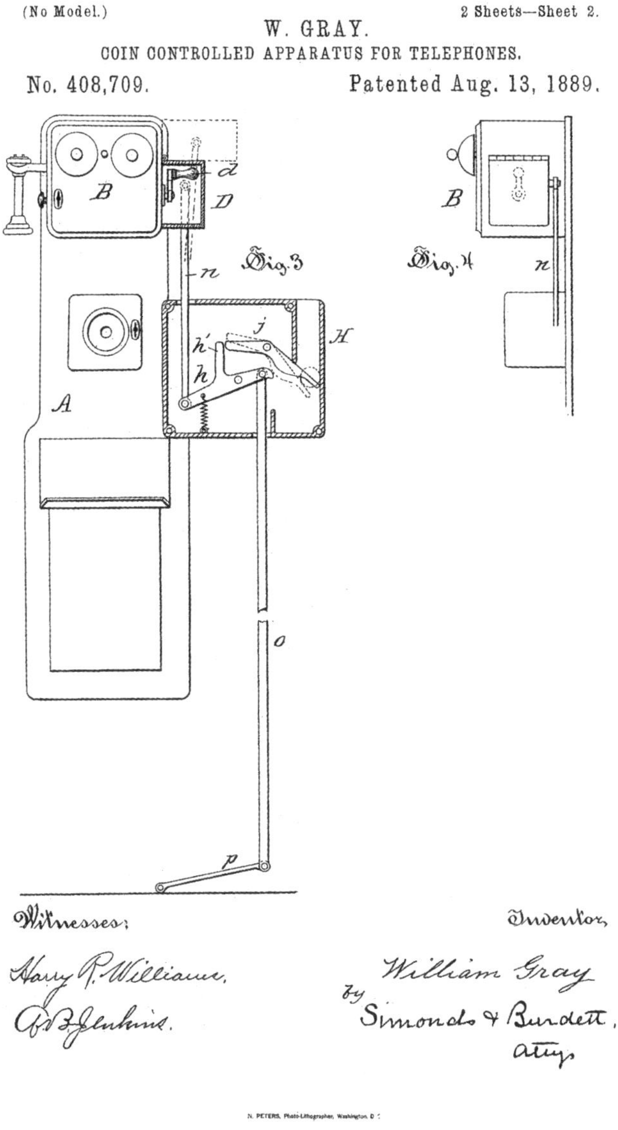

in Boston, women's skills proved superior to those of men, and the switchboard operator's profession became one of the most important areas of employment for women in several countries with a telephone network.[12]

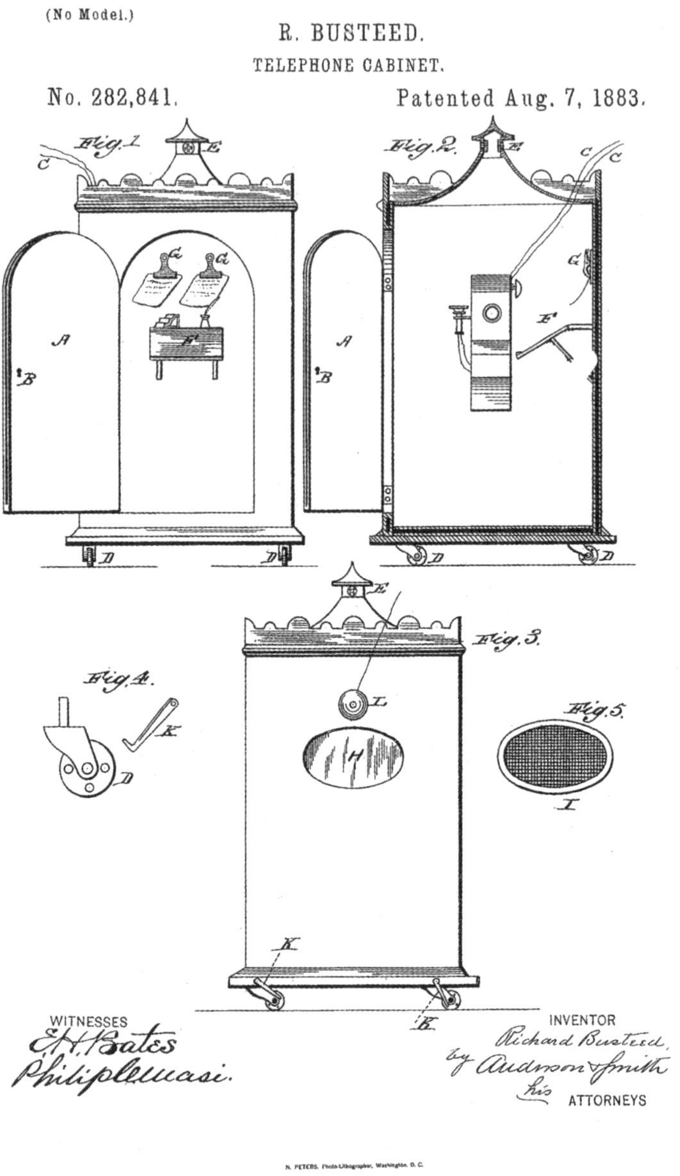

Richard Busteed, Telephone Cabinet, no. 282,841, patented August 7, 1883.

In France, automatic pay phones were introduced in two phases: in 1923, a first model called the F.W. Hall linked a money box to the telephone; and the following year the Hall / Roussotte integrated the money box with the telephone in the same housing.

Telephone booth at the Paris stock exchange, late nineteenth century. T10H6P. INTERFOTO / Alamy Stock Photo.

The latter has been described, in Simondian terms, as a form of "concretization" of the technical object, and as the perfection and new autonomy of the object through the fusion of two functionalities, which were previously juxtaposed.[13] These devices were always meant for use indoors, especially with the launch of the

The central telephone office on rue La Fayette in Paris, 1885. Cnum—Conservatoire numérique des Arts et Métiers.

Taxiphone company. It was not until the 1970s that France tried to catch up with other countries by increasing the number of outdoor public pay phones and housing them in telephone booths. In Italy, for example, the first outdoor metal and glass telephone booth had been installed in 1952 in Milan, in Piazza San Babila, by the concessionary operator Stipel.

After a first period of differentiation, made of wood or cast iron and of different shapes, in the United Kingdom the phone booth developed its strong iconic identity from 1921 right up to the 1980s, when Post Office took over the management. Also known as the Red Telephone Box, the public telephone kiosk became a national icon in the United Kingdom and its colonies (current and former). From the K1 to the K8—the K6 being the most common—a series of bright red color-coded kiosks, with the British crown symbol above the "telephone" sign, became an emblem of the country, dotting the territory.[14] As a result of a national competition in 1924, the architect Giles Gilbert Scott, author of the monumental Liverpool Cathedral, the Cambridge Library, the Battersea and Bankside electrical stations (the future Tate) designed the K2, and the rest of the models, up to K6.

AT&T advertising, 1940. N. W. Ayer Advertising Agency Records, Archives Center, National Museum of American History, Smithsonian Institution.

AT&T advertising, 1938. N. W. Ayer Advertising Agency Records, Archives Center, National Museum of American History, Smithsonian Institution.

Scott was known as an architect of large-scale projects. Curiously enough, he found inspiration for these famous national kiosks from an even more unexpected source: the mausoleum built by John Soane for his deceased wife—which would also contain the remains of his eldest son and subsequently his own—in the old St. Pancras Cemetery in London. The connection between a monument dedicated to the eternal and private memory of a single family and the creation of a public space dedicated to the modern use of instantaneous telecommunication is a striking one. It was most evident in the shape of the top of the cabin, which clearly refers to the pendentives and the tympanum of the flattened dome of the mausoleum. While the first versions of the booth were made of concrete (the K1 and the K3), the K2 and the next and more widespread models would be made of cast iron, until the much finer structure of the K6, an interpretation of the 1930s streamline style, were built for the jubilee of King George V, in 1935.

The British booths, like the pillar mail boxes, are always in red and have unparalleled levels of popular recognition. They provided a background for pop scenarios, including the opening scene of *A Hard Day's Night* with the Beatles (1964), in which

Giles Gilbert Scott, design for the British K2 telephone booth, competition 1924, production 1926. RIBA Collections.

three members of the group—George, Ringo, and John—emerge simultaneously from a triple phone booth. The back cover of *The Rise and Fall of Ziggy Stardust and the Spiders from Mars* (1972) features David Bowie as the titular alien messiah in a British phone booth on Heddon Street, London. Like historical monuments, the British telephone booths are cast in miniature as tourist souvenirs. They signal the widespread availability of a national public service, while becoming pieces of cult urban furniture, and even, nowadays, fetish objects.[15]

The telephone booth has also become an emblematic object of public service in developing countries and societies. It has entered the modern imaginary through popular culture, from comics to cinema. The reporter Clark Kent uses it whenever he needs a protected place in the heart of the city to transform himself into Superman. In Alfred Hitchcock's film *The Birds* (1963), Tippi Hedren sees the terrifying onslaught of birds from within a closed glass telephone booth, and in *Rosemary's Baby* (1968), Mia Farrow desperately looks for an escape over the phone as she feels spied upon from outside the booth. But the phone booth can also be the source of an anonymous call, untraceable, from any location in the city or territory. One can place a call, be called back, and leave without a trace. A number of thrillers, detective films, and road movies include a scene with a phone booth not only for its specific use as public, collective, and therefore anonymous property, but also for its secret and potentially dramatic use. For example, in the movie *Duel* (Steven Spielberg, 1971), and *Phone Booth*, a film by Joel Schumacher (2002), the latter's plot entirely built on the suspense created by a call received from a phone booth. This cinematic exploitation of the phone booth in particular and its infrastructure in general contains a dramaturgical potential linked to spatio-temporal narration, which is expressed in the material power of the booth's presence and repetition throughout the territory, a form of "natural telegeny," and observed through a process of spontaneous mythification.[16]

It turns out that the phone booth also represents a particular kind of space as defined by law. An interesting anecdote reveals how the public pay phone, and the telephone booth as a place, came to acquire the legal status of private space when occupied by a user. In 1967, the FBI discovered that a man named Charles Katz was placing illegal bets in Miami and Boston from a pay phone on Sunset Boulevard in Los Angeles.[17] Through an agreement with

Excerpt from the film *A Hard Day's Night* (directed by Richard Lester) released with the Beatles' eponymous album, 1964—G656P8. PA Images / Alamy Stock Photo.

the phone company, the booth's telephone was tapped, and this provided the evidence for Katz's arrest. In his trial, Katz argued that the wiretap constituted an invasion of privacy. The Supreme Court, referring to the Fourth Amendment, which "protects people, not places"—and thus the person making the phone call, even if from inside a public phone booth—eventually agreed with him: judging that someone doing business from a phone booth should have a "reasonable expectation of privacy." From that point on, no pay phone could be tapped without a search warrant, making it a popular venue for criminal activity that peaked in the 1990s and prompted some U.S. municipalities to establish restrictions on their use. Pay phone companies, recognizing that their product was a common tool for drug dealers, gamblers, pimps, and hustlers especially given privacy laws that allowed users to remain anonymous, agreed to limit the service to outgoing calls. In some

Back cover of David Bowie's album, *The Rise and Fall of Ziggy Stardust and the Spiders from Mars*, 1972. © The David Bowie Archive. ® Photo by Brian Ward.

cases, booths were banned from the city altogether, as in Chicago (as well as in Philadelphia and Kansas City) by a 1992 decision.[18] Other municipalities eventually defined specific locations in the city where booths could be used, as well as other locations where booths were to be removed, such as in Los Angeles, New Orleans, and Dover, Delaware—particularly in the vicinity of liquor stores or in particular areas of public housing, thereby disrupting access to this public service. However, this form of temporary privatization of the public object at the moment of its use becomes mostly significant in this context, and it further emphasizes this suspended condition, in the image of an isolated bubble, not only for simple acoustic reasons, but also in the face of the law and the definition of public space and utility.

With the phone booth, telephone service becomes an object, a standardized piece of urban furniture; we can speak of

Clark Kent transforms into Superman in a phone booth. Comic strip, 1942.

microarchitecture here, as the construction makes it possible to house a function, to create a confined and distinct space. Acoustics is the parameter that is mainly controlled, to create a quiet, isolated, and intimate space. But light and climatic conditions are also often taken into consideration in order to create a place of comfort for the duration of a telephone conversation: with transparent glass and artificial lighting for sufficient light, as well as physical shelter to protect from the sun, wind, and rain. If modernity made the functions of architecture legible through its technical specialization, the telephone booth is in a way a scaled-down version of machine architecture, a device that requires a specific container to house its own function. Apart from the English telephone booth and a few other particularly expressive examples, the form and appearance of the box is usually reduced to a great simplicity, and it eventually acquires an aesthetic code through its temporal and geographical classification: the choice of materials used, the design of its more or less geometric surfaces; any logos, advertisements, or signs of belonging; the various models of telephone introduced, as well as ancillary accessories (telephone bulletin, shelf, lamp, token dispensers, telephone cards, and so on).

Previously marked by access points, the ramified telephone network is in the process of becoming obsolete. Indeed, the digital era opens the way to a portable, individual, and private system that does away with the public mapping of the network. Beginning in the 1980s in most Western countries, the process of privatizing public networks has taken place in parallel with the development of mobile telecommunications, mainly for economic reasons linked to the competitive telecommunications market. The two apparently separate phenomena established the basis for a radical transformation of telephone communication methods. Thus, the phone booth represents the last bastion of public service in all its architectural ambiguity: a closed, intimate, and private space in the middle of a shared national network and so-called public territorial space. It also indicates that an underground or aerial system of wires and cables connects this precise point to a diffuse and common network, even in a particularly isolated corner of the territory.

Since the advent of mobile phones, the concept of the network has evolved considerably. While nowadays "having a network" means that we are covered by electromagnetic waves allowing us to access wireless communication, we don't necessarily know where physical infrastructures are and who owns the transmission channel, the relay antenna, or the communication transmission circuit. Our private cell phone accesses the network more and more freely: the telephones that were once fixed points of the territory and were embodied in the public domain by telephone booths and fixed devices have now become individual and mobile devices. Strictly speaking, we no longer have the space of comfort created by the microarchitecture of the booths, and we recreate it by seeking the conditions of an aside, or by means of headphones incorporated on or in our auditory canals, or by a hand that directs the voice toward the microphone. Visually, from a map reproducing a network, intersections, nodes, and points, we move to a representation of clouds. These vary in density, and they move in space according to the time and levels of activity.

10 *Ubiquity: Finder of Men*

In 1922, László Moholy-Nagy commissioned a series of five enamel paintings from a Berlin sign factory, to be made according to instructions he had given over the telephone. In so doing, he called into question the very foundations of artistic creation.

In keeping with his approach to the disembodiment of the creative gesture, Moholy-Nagy rejected any claim to authorship, thus subscribing to the spirit of modernist culture just as much as to the era of mechanical reproduction. "In 1922 I ordered by telephone from a sign factory five paintings in porcelain enamel. I had the factory's color chart before me and I sketched my paintings on graph paper. At the other end of the telephone the factory supervisor had the same kind of paper, divided into squares. He took down the dictated shapes in the correct position. (It was like playing chess by correspondence)."[1]

Further on, Nagy wrote: "I was not at all afraid of losing the 'personal touch,' so highly valued in previous painting. On the contrary, I even gave up signing my paintings. I put numbers and letters with the necessary data on the back of the canvas, as if they were cars, airplanes, or other industrial products."[2] The loss of what he called the "personal touch" is in fact confirmed on two levels: on the one hand, the production of the painting is an industrial process devoid of any manual intervention on the part of the artist; on the other hand, the instructions necessary for its

László Moholy-Nagy, EM2 (*Telephone Picture*), 1923. Museum of Modern Art, New York. Gift of Philip Johnson in memory of Sibyl Moholy-Nagy. Museum of Modern Art (MoMA), New York, USA.

realization are provided to the factory by telephone, without any plan being sent or any meeting taking place on the spot between client and company. In this way, László Moholy-Nagy submits the work of art to the challenge of the modern telecommunications system, engaging in direct exchange between individuals while at the same time abolishing human contact and the visual counterpart of a meeting. Moreover, the titles of the original works—now commonly referred to as Telephone Paintings—emphasized the material and manufactured nature of the objects ("The Enamel Paintings") as well the idea of anonymity and series. The paintings were labeled "EM" and numbered from EM1 to EM5. Reading Moholy-Nagy's personal account, *Abstract of an Artist*, we learn about the creative process of this work in relation to telecommunication, even if the testimony of his first wife Lucia Moholy-Nagy casts doubt on its veracity.[3] But it is not so much the highly conceptual nature of his process, or the medium employed that gives this work its particular significance, but rather the way in which it combines art, technology, science, and everyday practices.[4] A year later, in 1923, Moholy-Nagy replaced Johannes Itten as professor of the preliminary course at the Bauhaus. That year marked a decisive turning point in the school's mode of art education: it moved away from romantic and spiritual conceptions to a much more pragmatic approach that integrated design and industry.

Thirty-six years later, in 1968, the Museum of Contemporary Art in Chicago drew on Moholy-Nagy's experience to launch the exhibition *Art by Telephone* at the initiative of curator Jan van der Marck. It opened the following year and brought together works of thirty-seven American and European artists, created on site from instructions transmitted by telephone.[5] A vinyl catalog featuring their recordings documented this moment of creation by delegation, which this time focused on the conceptual dimension of the approach.[6]

> Conceptual art as documented, recorded, manufactured, or preformed in *Art by Telephone* is a further step toward the syncretism of the literary, plastic and performing arts which characterizes the 1960s. The term generally applies to those new forms art which seem to favor intellectual premises over visual result. Those artists who have responded to the challenge to this exhibition share certain basic premises despite divergence in expression. They want to get away

> from the interpretation of art as specific, handcrafted, precious object. They value process over product and experience over possession.[7]

Here again the telephone is used to express the process of disembodiment of the work of art. It is not so much the creative gesture and the manual skill of the artist that counts as the clarity of an idea and the precision of the words communicated by electrical means. The curator, at the other end of the line, then takes care of the concrete realization of the work. This idea of transmission at a distance between artist and curator inspired other projects such as the version of the *Art by Telephone* exhibition presented in 2010 at the contemporary art fair Access & Paradox in Paris: the material necessary for the installation was sent by mail to Cécile B. Evans and Rebecca Lamarche-Vadel, with the instructions for its realization provided during a videoconference. The modes of communication become more complex—sound, visual and postal messages—but the idea of a work realized by delegation persists. In the same vein, curators Sébastien Pluot and Fabien Vallos conceived a project to reactivate the 1969 exhibition, in November 2012, unfolding in six different locations in France and the United States.[8]

In these examples, on the one hand, the telephone allows rethinking the figure of the artist in terms of a sound machine, an electric voice, or as one of the extremities of a network, playing ambiguously on the loss of individuality, the process of anonymous production and thus the reproducibility of the work. On the other hand, the effect of disincarnation questioned by these approaches is also explored in a critical or ironic way, by underlining the alienating dimension of the telephone conversation. The fact of talking to a machine, of arguing and loving by interposed electric wires, constitutes both a real revolution as well as a change in the system of the social relations that leads to reflections.

If one admits, to borrow from Marshall McLuhan, that the artist is a kind of radar, an early warning system of change, anticipating technological and social transformations generated by the new media, these various pieces reveal several ways of looking at the telephone system beyond the more retinal approach put forward at the beginning of this book.[9] From a formal, imaginative, sublime, and allegorical inspiration of the network and its physical materiality, we pass to works that question more the mode of use of the telephone and its inevitable implications for society.

In 1966, during a period of strong conceptualization of art work, the telephone finally became a real artistic medium showcased at *9 Evenings: Theatre and Engineering*, a series of performances at the 69th Regiment Armory in New York. Produced by an association of ten artists[10] and thirty Bell Laboratories engineers, and organized by Robert Rauschenberg and Billy Klüver—himself a former Bell engineer—*9 Evenings* was the first in a series of projects known as Experiments in Art and Technology. These experiments demonstrated that it was no longer simply a question of using the usual functions of the telephone, but rather of putting technologies conceived by telecommunications engineering at the service of art (dance, music, and theater).

Thus, for John Cage's performance entitled *Variations VII*, Billy Klüver obtained from New York Telephone the equipment necessary for the installation of ten telephone lines left open to capture sounds from different spaces: a restaurant on Fourteenth Street, the aviary at the Bronx Zoo, a power station, a room for holding lost dogs, a garbage dump, a press room of the *New York Times*, even Merce Cunningham's studio and Terry Riley's home, and other microphones were connected to the stage.[11]

Made during an intense period of infatuation with electricity's power, this work focuses on the potential for transmission and amplification of sound offered by technology. Unlike the works created "by telephone," here the machine, its system, and its network constitute the artistic medium itself, like the brush and the canvas, or the enamel paint. The work itself is spatialized. It becomes a museum and an exhibition, a concert and music hall, as the container of the 69th Regiment Armory fades away in the presence of a staged network and a recreated physical and sound environment.

These different works insist on the ubiquitous character of telecommunication—the possibility of being in several places at a given time, at least by sound. Our trajectory of following the line ends with the figure of the user—the last (and first) point of arrival (and departure). The way in which the voice, with all its nuances, is technically transmitted through the airwaves is again easily described in Hopkins's 1899 manual:

> The pitch of the sound depends upon the period or frequency of vibrations of the fundamental note, and would be the same were all the overtones destroyed. Nor is the pitch

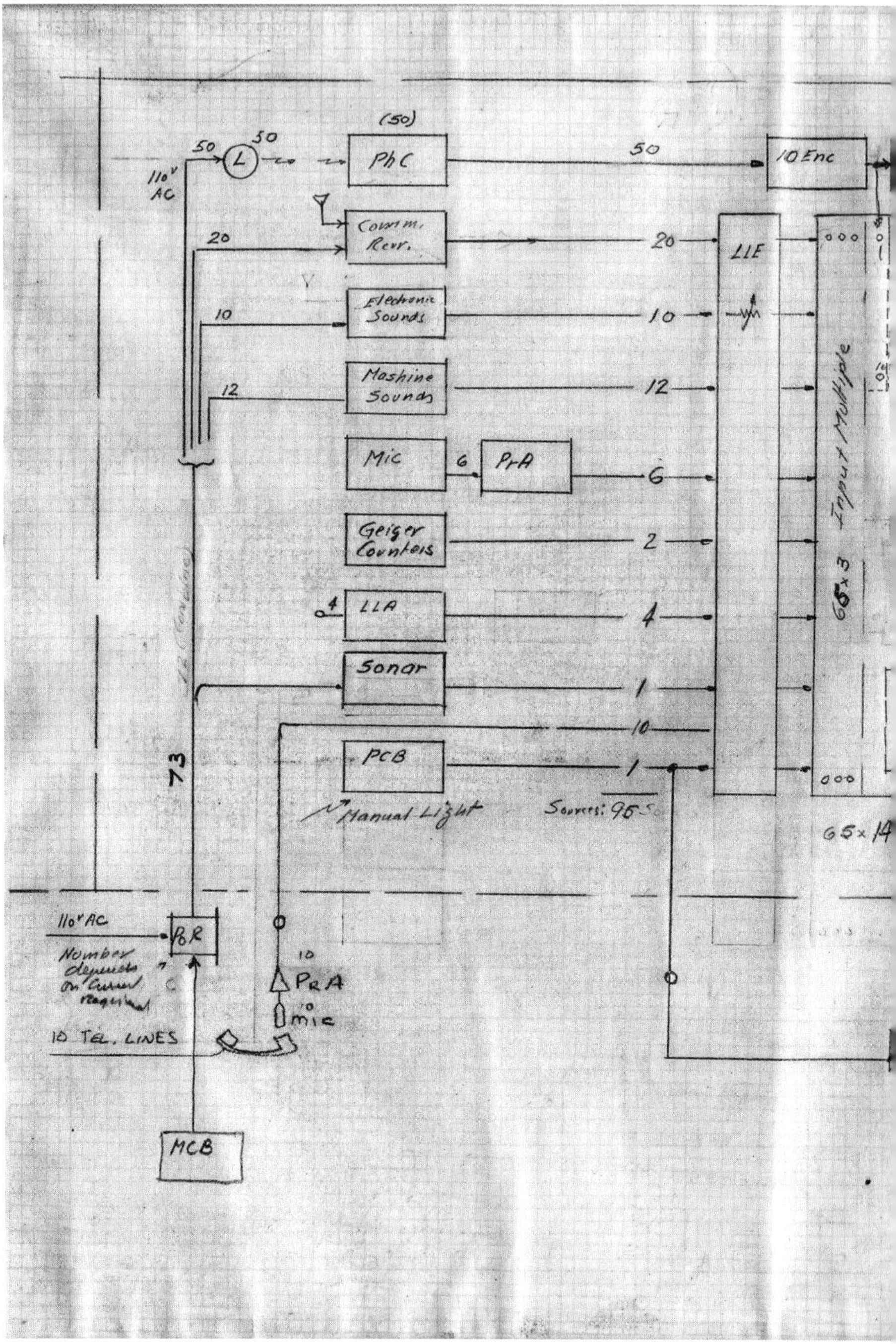

Herb J. Schneider, diagram for *Variations VII* by John Cage, *9 Evenings: Theatre and Engineering* (1966). Courtesy: Experiments in Art and Technology and Robert Rauschenberg Foundation.

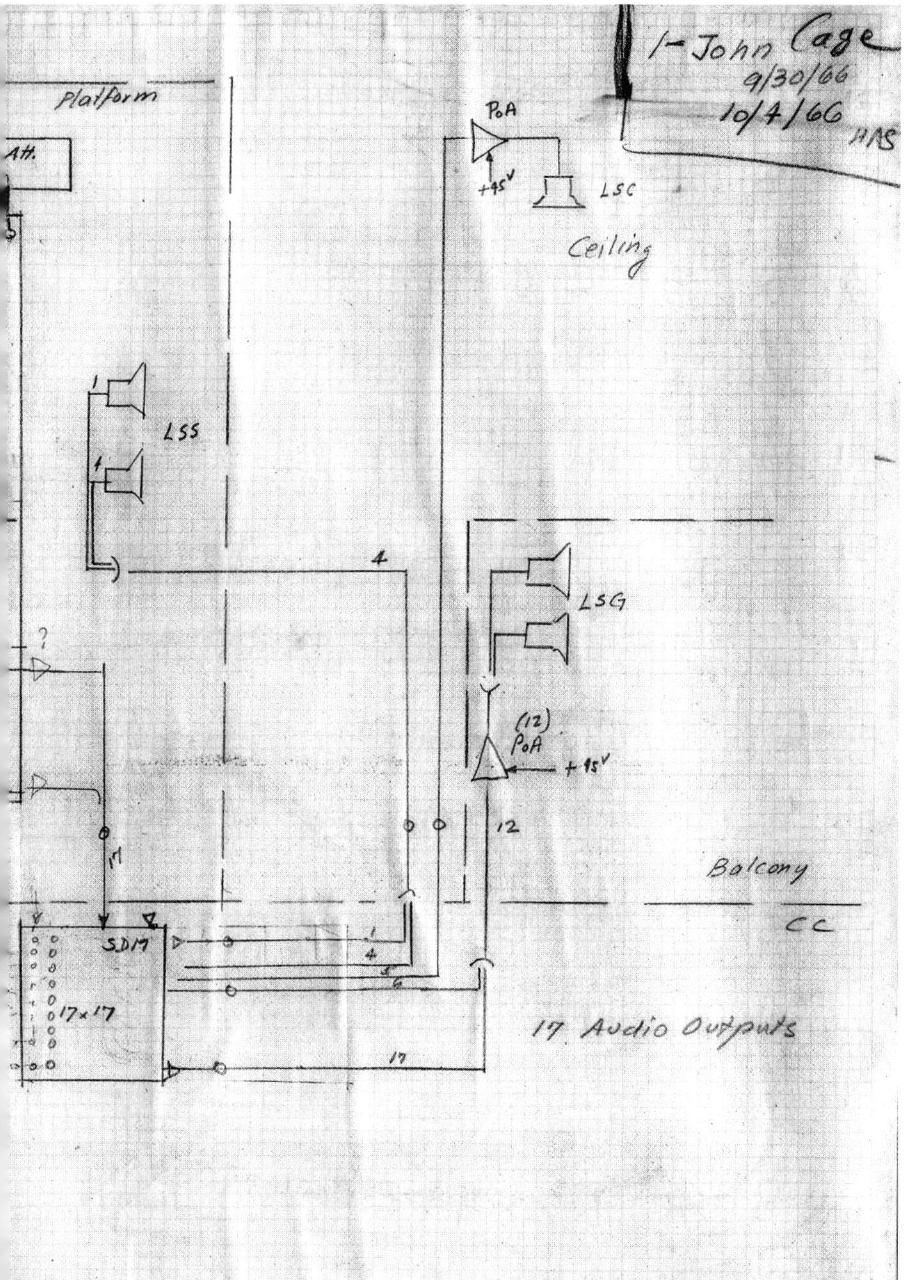

1- John Cage
9/30/66
10/4/66
HAS
Platform
Att.
PoA
+15V
LSC
Ceiling
LSS
4
LSG
?
(12)
PoA
+15V
12
17
Balcony
CC
SDM
17×17
1
4
5
6
17
17 Audio Outputs

> at all dependent on the form of the wave. The form of the wave determines the quality or character; and the resultant wave produced by the voice, including all the overtones, is generally of a very complex form. It is by the quality or character that we are able to recognize individual sounds. Clearness depends upon the preservation of the form of the wave, and loudness upon the amplitude of the vibrations."[12]

The entire difficulty for the telephone, Hopkins states, is determining how to regain the complex starting properties of the sound wave, at the end of its journey through the power lines. The extent of loss of clarity and intensity of the signal, by transmission through alternating current, would be the deciding factor determining its function and success.

For the ever-effective corporate rhetoric, the telephone becomes a "finder of men," capable of isolating "the man in the multitude," like an electric flashlight that illuminates a target. This ability to focus, to isolate an element in a multitude by the added power of technology, echoes Walter Benjamin's definition of modern man and his need to find a refuge in the midst of the crowd. Celebrated as products of the machine age, telecommunication also represents a possible form of defense, of protection, a moment of reconciliation with the individual and the small scale of the human at the same time. While these multiscalar systems cross cities, regions, countries, oceans, and continents, the networks of telecommunications, this great human enterprise, nonetheless address the individual and his or her intimate space.

The objective of this book, as in all these different artistic explorations, has been to search for a new spectrum in which to view the modernization and mechanization of the landscape, through the introduction of telecommunications infrastructures. This spectrum is made up of objects that are often relegated to the background. In a reading of history centered on the "great men" of the modern era, those famous actors, whether they be inventors, entrepreneurs, architects, urban planners, and so on—"men," in the specific sense of the term, because as we know that in the field of technology, as in the domains of engineering and architecture, a male dominance has influenced the historical course of the last century, as well as the retrospective reading of this period.

Thus, this book is situated in a different historiographical tradition, one that takes into account figures and technical elements

AT&T advertising, 1910. N. W. Ayer Advertising Agency Records, Archives Center, National Museum of American History, Smithsonian Institution.

AT&T advertising, 1915. N. W. Ayer Advertising Agency Records, Archives Center, National Museum of American History, Smithsonian Institution.

that are often considered secondary, but without which that same history could not have existed. If Brenda Maddox advances the hypothesis that the invention of the telephone was inspired by the existence of two little-known but central women in history—Alexander Graham Bell's mother and wife, who were both deaf—these different texts also try to change the focus of the usual narratives.[13] What happens if we look at the history of the telephone through close observation of its system, its network, its wires, cables, and plugs? What perception of the telecommunications era can we have if we consider the popular press, corporate propaganda, or legislation as parts of a corpus? What is the gap between these forms of reception and the great narratives of architects, theorists, and the artists' productions of the same period?

Despite the variety of themes, scales, historical and geographical contexts, this work has allowed us to explore transversal theses. When we look at the architecture from its relation to the infrastructures of communication, sound, and information, we leave definitively any static vision of the objects. Architecture becomes a medium of passage or a temporal terminus of a highly articulated system of transmission, a movement very often schematized by the figure of the line. It is not only an interpretation of the technical function, but also the legal status, the reception of the users that demonstrates the impossibility of isolating the architecture from the rest of the network. Thus, even when we have tried to limit our observation to precise objects composing a transmission system—the conduits, the poles, the outlets, or the telephone booth—we are constantly carried by a tension, a movement that pushes us to think instead in terms of series and successions. If telecommunications systems have contributed to rethinking the organization of spaces, architecture, the city, and the territory, their development and constitution are not isolated but result from the conjunction of several material and cultural factors that allow them to exist.[14] These factors include the entire modern infrastructure network. Social, economic, and political conditions have created a fertile ground for their use and adoption in everyday life. As we have seen, their horizontal as well as vertical development has not actually transformed the ontological dimensions of time and space, contrary to the claims of corporate publicity, but has certainly facilitated the establishment of new relationships.

The assimilation of these networks has opened up as many real possibilities as prophetic phantasmagoria in literature, art, architecture, and popular culture. In general, this boundary between the real and the imaginary has remained very blurred, because the

Videophone, 1920. Peter Weiss Collection.

shift from one dimension to the other could still happen, thanks to the unpredictable possibilities of science and technology in a given social and cultural context.

Thus, if reality has nourished fiction, the reverse is also true and the iconographic richness that results from this encounter is a valuable illustration of how this vast modern, mechanical universe overcomes several antithetical categories of the preindustrial tradition, such as nature and culture,[15] countryside and city, sublime and beautiful, architecture and network, private and public, object and symbol, object and thing.[16]

Acknowledgments

This book is the result of many years of archival research and personal encounters. I would like to express my gratitude to all the people who helped me in various ways through different institutions: the National Museum of American History at the Smithsonian Institution, the Jerome and Dorothy Lemelson Center for the Study of Invention and Innovation, the AT&T Archives and History Center at Warren (New Jersey), the Andrew W. Mellon Foundation, the Department of Art History and Communication Studies at McGill University, the Canadian Center for Architecture, the Bibliothèque historique des postes et télécommunications and the Bibliothèque nationale de France, both in Paris.

Thank you to the many colleagues and friends with whom I have been able to share and develop ideas: Marc Angélil, Marco Assennato, Darin Barney, Pierre Caye, Adrian Forty, Moritz Gleich, Vittorio Magnago Lampugnani, Arthur P. Molella, Antoine Picon, Alessandra Ponte, Marc Raboy, Jonathan Sterne, Will Straw, Nicolas Tixier, Melissa van Drie.

The making of the book is also the fruit of many extraordinary efforts and skills. The original French version was published by MétisPresses, in Geneva, which generously shared its files with the MIT Press. For the English translation, I would like to thank Christian Hubert for his commitment and passion and Emma Brown for brilliantly managing illustrations and permissions. I am indebted to the expert editorial team at the MIT Press. Finally, my deepest gratitude goes to Thomas Weaver for his unique and visionary way of making books.

Finally, a fond thought goes to Sofia, Olmo, Lucio, and Vito, who came into the world during these years of research. I dedicate this book to Laurent, with whom I share it all.

Notes

Preface

1. Ithiel de Sola Pool, *The Social Impact of the Telephone* (Cambridge, MA: MIT Press, 1977).

2. See, among others, *The Information Age: Economy, Society and Culture* trilogy by Manuel Castells, published between 1996 and 1998 (Oxford: Blackwell); Stephen Graham and Simon Marvin, *Splintering Urbanism: Networked Infrastructures, Technological Mobilities and the Urban Condition* (London / New York: Routledge, 2001); as well as Keller Easterling, *Extrastatecraft: The Power of Infrastructure Space* (London / New York: Verso, 2014); Gretchen Bakke, *The Grid: The Fraying Wires Between Americans and Our Energy Future* (New York / London, Bloomsbury, 2016); Shannon Mattern, *Code and Clay, Data and Dirt: Five Thousand Years of Urban Media* (Minneapolis: University of Minnesota Press, 2017).

3. Emily Thompson, *The Soundscape of Modernity: Architectural Acoustics and the Culture of Listening in America, 1900–1933* (Cambridge, MA: MIT Press, 2002).

4. A tradition that stems from the seminal work of Sigfried Giedion, *Mechanization Takes Command: A Contribution to Anonymous History* (New York: Oxford University Press, 1948); and Reyner Banham, *The Architecture of the Well-Tempered Environment* (London: Architectural Press, 1969).

5. Stanislaus von Moos, *Esthétique industrielle* (Disentis: Pro Helvetia/Editions Desertina, 1992).

6. Franz Graf and Giulia Marino, eds., *Les dispositifs du confort dans l'architecture du XXe siècle: connaissance et stratégie de sauvegarde* (Lausanne: Presses polytechniques et universitaires romandes, 2016).

7. Eduardo Prieto, "La culture du bien-être. Les poétiques du confort dans l'architecture du XIXe et du XXe siècle," in Graf and Marino, *Les dispositifs du confort dans l'architecture du XXe siècle*, 41–58.

8. Von Moos, *Esthétique industrielle*, 1.

9. Georges Simondon, *Du mode d'existence des objets techniques* (Paris: Aubier, [1958] 2015).

10. Simondon, *Du mode d'existence des objets techniques*, 185.

11. Simondon, 186–187.

Chapter 1

1. Manuel Castells, *The Rise of the Network Society*, vol. 1 of *The Information Age: Economy, Society and Culture* (Malden/Oxford: Blackwell, 1996); David Harvey, *Spaces of Capital: Towards a Critical Geography* (Edinburgh: Edinburgh University Press, 2001).

2. Anna Tsing, *Friction: An Ethnography of Global Connection* (Princeton: Princeton University Press, 2005), 4; Rania Ghosn, "Territories of Oil: The Trans-Arabian Pipeline," in *The Arab City. Architecture and Representation*, ed. Amale Andraos and Nora Akawi (New York: Columbia Books on Architecture and the City, 2016), 164–175.

3. Martin Heidegger, "The Thing," in *Poetry, Language, Thought*, trans. Albert Hofstadter (New York: Harper and Row, [1951] 1971), 165–186.

4. Darin Barney, "Pipelines," in *Fueling Culture: Politics, History, Energy*, ed. Imre Szeman, Jennifer Wenzelt, and Patsy Yaeger (New York: Fordham University Press, 2017), 267–270. Citations on 268 and 269.

5. William J. Hopkins, *Telephone Lines and Their Properties* (New York: Longmans, Green, and Co., 1899). See also C. H. Bristol, *The Western Union Telegraph Company: Rules for Construction and Repairs* (New York: The Western Union Telegraph Company, 1905).

6. Hopkins, *Telephone Lines and Their Properties*, 13.

7. Ward v. Atlantic & Pacific Telegraph Company, New York, 1877. Edward Q. Keasbey, *The Law of Electric Wires in Streets and Highways* (Chicago: Callaghan and Company, 1900). Reprinted by Biblio Bazaar, 2009, 264.

8. John R. Stilgoe, *Outside Lies Magic* (New York: Walker & Company, 1998), 21.

9. Stilgoe, *Outside Lies Magic*, 22.

10. Bakke, *The Grid*.

11. Jean Baudrillard, *Le système des objets* (Paris: Gallimard, 1968).

12. Marshall H. McLuhan, *D'oeil à Oreille*, trans. Derrick de Kerckhove (Montréal: Éditions Hartubise HMH, 1977), 36–37.

13. Alessandra Ponte, "Journey to the North," in *The House of Light and Entropy* (London: Architectural Association, 2014), 135–163.

Chapter 2

1. David E. Nye, *American Technological Sublime* (Cambridge, MA: MIT Press, 1994).

2. Derek P. McCormack, *Atmospheric Things: On the Allure of Elemental Envelopment* (Durham, NC / London: Duke University Press, 2018). See esp. the chapter "Sounding," 121–143.

3. Jane Bennett, *Vibrant Matter: A Political Ecology of Things*

(Durham, NC: Duke University Press, 2010).

4. These different descriptions of the "telegraph harp" are found in *The Writings of Henry David Thoreau, Journal III, 1851–1852* (Boston: Houghton Mifflin Co., 1906). The notes were taken by the author during his stay of two years, two months, and two days in a cabin in the woods surrounding Walden Pond.

5. Catherine Maumi, *Usonia ou le mythe de la ville-nature américaine* (Paris: éd. de la Villette, 2008), 31.

6. Leo Marx, *The Machine in the Garden: Technology and the Pastoral Ideal in America* (London: Oxford University Press, 1964). See also Barbara Novak, *Nature and Culture: American Landscape and Painting* (London: Thames & Hudson, 1980), see chap. 8, "Man's Traces: Axe, Train, Figure," 157–199.

7. Nye, *American Technological Sublime*.

8. Susan Glickman, *The Picturesque and the Sublime: A Poetics of the Canadian Landscape* (Montreal / Kingston: McGill-Queen's University Press, 1998), ix, 9–10.

9. The image was a mockup for a chromolithograph commissioned by George Crofutt, the American publisher of a travel magazine that was widely popular at the time. Richard R. John, *Network Nation: Inventing American Telecommunications* (Cambridge / London: The Belknap Press of Harvard University Press, 2010), 111.

10. For a complete description of the image, see Martha A. Sandweiss, *Picturing U.S. History*, accessed November 13, 2019, https://picturinghistory.gc.cuny.edu/john-gast-american-progress-1872.

11. Alain Beltran and Patrice A. Carré, *La fée et la servante: la société française face à l'électricité, XIXe-XXe siècles* (Paris: Éditions Belin, 1991); Martine Marin, "Femmes et société: le travail ménager (1919–1939)," PhD diss., Université Paris-Diderot Paris 7, 1984 (Michelle Perrot, thesis advisor). See also Graeme Gooday, *Domesticating Electricity: Technology, Uncertainty and Gender, 1880–1914* (London: Pickering & Chatto, 2008).

12. Images and quotations from the official announcements on electricity. Michelle Perrot, "Histoire de la condition féminine et histoire de l'électricité," in *Électricité dans l'histoire: problèmes et méthodes*, conference proceedings of the Association pour l'histoire de l'électricité en France, Paris, October 11–13, 1983, 176.

13. Daniel Webster, *Mr. Webster's address at the laying of the cornerstone of the addition to the Capitol, July 4th, 1851* (Washington: Gideon and Co. Printers, 1851), 18; François Jarrige, *Techno-critiques: du refus des machines à la contestation des technosciences* (Paris: Éditions La Découverte, [2014] 2016), 126.

14. Roger Boulet, ed., *Vistas: Artists on the Canadian Pacific Railway* (Calgary: Glenbow Museum, 2009), 108.

15. Northrop Frye, "The Canadian Scene: Explorers and Observers," in *Canadian Landscape Painting, 1670–1930: The Artist and the Land*, ed. Robert Hamilton Hubbard (Madison: University of Wisconsin Press, 1973), 1–29.

16. Harold A. Innis, *Staples, Markets, and Cultural Change: Selected Essays* (Montreal / Kingston: McGill-Queen's University Press, 1995). See esp. "Transportation as a Factor in Canadian Economic History," 123–138. See Jody Berland, "Space at the Margins," in *North of Empire: Essays on the Cultural Technologies of Space* (Durham, NC, Duke University Press, 2009), 65–97.

17. Novak, *Nature and Culture*, 166.

18. Jarrige, *Techno-critiques*, 177.

19. Edmond B. Ogle, *Long Distance Please: The Story of the Trans Canada Telephone System* (Toronto: Collins, 1979), 41.

20. Roland Marchand, *Advertising the American Dream: Making Way for Modernity, 1920–1940* (Berkeley: University of California Press, 1985) and John, *Network Nation*.

21. Tim Ingold, *Une brève histoire des lignes*, trans. Sophie Renaut (Brussels: Zones sensibles, [2007] 2013), 222.

22. Paul Virilio, *Ce qui arrive* (Paris: Éditions Galilée, 2002).

23. The laws governing the installation of networks were defined by each municipality. See "Underground Wires," in Keasbey, *The Law of Electric Wires in Streets and Highways*, 84–107.

24. John Brooks, *Telephone: The First Hundred Years* (New York: Harper & Row Publishers, 1975), 87.

25. Elected governor of Tokyo in 2016, Yuriko Koi set a goal for the 2020 Olympics to bury the capital's six million power poles and cables.

26. Andres Lepic, "Mies and Photomontage, 1910–38," in *Mies in Berlin*, ed. Terence Riley and Barry Bergdoll (New York: Museum of Modern Art, 2001), 326.

27. Adrian Sudhalter, "Friedrichstrasse: The Contexts of an Image, 1922–1924," in *Mies van der Rohe: Montage Collage*, ed. Andreas Beitin, Wolf Eiermann,

and Brigitte Franzen (London: Koenig, 2017), 68–85.

28. Frank Lloyd Wright, *The Living City* (New York: Horizon Press, 1958), 136.

29. Wright, *The Living City*.

30. Mathew Aitchison, "The Boyd Ultimatum," *AA Files* 66 (2013): 59–67.

31. Robin Boyd, *The Australian Ugliness* (Melbourne: Cheshire, [1960] 1963), 38–39, cited in Aitchison, "The Boyd Ultimatum."

32. Aitchison, "The Boyd Ultimatum."

33. John B. Jackson, "Other-Directed Houses," *Landscape* (1956), published in *Landscape in Sight: Looking at America*, ed. Helen Lefkowitz Horowitz (New Haven: Yale University Press, 1997), 185–186.

34. Robert Venturi, Denise Scott Brown, and Steven Izenour, *Learning from Las Vegas: The Forgotten Symbolism of Architectural Form* (Cambridge, MA: MIT Press, [1972] 2001).

35. François Béguin, "La quincaillerie du nouveau monde," *Le genre humain*, no. 34 (1999): 149–162.

36. Béguin, "La quincaillerie du nouveau monde," 149.

37. Béguin, 149.

38. Béguin, 149.

39. Gildas Baudez and François Béguin, *Critique du paysage de l'énergie* (Paris: Roux-Bauer and Associates, 1980), 31.

40. Florence Bancaud, "L'esthétique du laid, de Hegel à Rosenkranz. Une 'esthétique de la résistance' ou de la résignation aux 'arts qui ne sont plus beaux'?," *Études Germaniques* 256, no. 4 (2009): 900.

41. Bancaud, "L'esthétique du laid, de Hegel à Rosenkranz," 900.

42. Karl Rosenkranz, *Aesthetics of Ugliness: A Critical Edition* (1853), edited and translated by Andrei Pop and Mechtild Widrich (New York: Bloomsbury, 2015).

43. Rosenkranz, *Aesthetics of Ugliness*, 31.

44. Rosenkranz, 38.

45. Rosenkranz, 59–60.

46. Georges Farhat, "L'urbain comme paysage infrastructurel: Le val de Bièvre," in *La métropole des infrastructures*, ed. Claude Prelorenzo and Dominique Rouillard (Paris: Éditions Picard, 2009), 268–281.

47. Oral declaration regarding the painting, in Antoine Baudin, "Les constructeurs de Fernand Léger: De la place Smolensk à la vallée de Chevreuse, ou quelques tribulations d'un chantier pictural," *Matières*, no. 2 (1998): 71. See also Alexis Guilbert, "Très Haute Tension sur l'Architecture: L'implication des architectes dans la conception de pylônes électriques," Master 2 thesis, ENSA Paris Malaquais, January 2013, 38 (Dominique Rouillard and Nathalie Chabiland, thesis advisors).

Chapter 3

1. Grandville, *Un autre monde: Transformations, visions, incarnations, ascensions, locomotives, explorations, pérégrinations; explorations, excursions; stations, cosmogonies, fantasmagories, rêveries, folâtreries, facéties, lubies, métamorphoses, zoomorphoses, lithomorphoses, métempsycoses, apothéoses et autres choses* (Paris: H. Fournier, 1844).

2. Catherine Dunlop Mackenzie, *Alexander Graham Bell: The Man Who Contracted Space* (Boston / New York: Houghton Mifflin Company, 1928).

3. On the neighborhood metaphor, see John, *Network Nation*, 5–23.

4. On the perception of time and space in the nineteenth century, see Carolyn Marvin, "Annihilating Space, Time, and Difference," in *When Old Technologies Were New: Thinking About Electric Communication in the Late Nineteenth Century* (New York: Oxford University Press, 1988), 191–235. On transportation, see Wolfgang Schivelbusch, *The Railway Journey: The Industrialization of Time and Space in the 19th Century* (Berkeley: University of California Press, 1986), especially "Railroad Space and Railroad Time," 33–44.

5. Schivelbusch, *The Railway Journey*, 38.

6. Patrick Geddes, *Cities in Evolution: An Introduction to the Town Planning Movement and to the Study of Civics* (London: Williams & Norgate, 1915). See also Pierre Chabard, "Cities in Evolution (1915): The Writing and the Readings of an Intempestive Classic," in *Manifestoes and Transformations in the Early Modernist City*, ed. Christian Hermansen (London: Ashgate, 2010), 149–162. On Geddes and his theories of the city, see Volker M. Welter, *Biopolis: Patrick Geddes and the City Life* (Cambridge, MA: MIT Press, 2002).

7. Patrick Geddes, "Civics: As Applied Sociology. Part I," in *Sociological Papers 1904* (London: Macmillan, 1905). In 1923, invited by Mumford to give a series of lectures at The New School for Social Research, Geddes returned to the subject: "The Valley Plan of Civilization (The Third of the Talks from the Outlook Tower)," *Survey Graphic* (June 1925): 288–290, 322–323, 325, republished in Geddes, *Cities in Evolution*, new and revised edition (London: Williams and Norgate, [1915] 1949), 164–167. See also Jaqueline Tyrwhitt,

"The Valley Section: Patrick Geddes's World Image," *Journal of the Town Planning Institute*, no. 37 (January 1951): 61–66.

8. *Report of the New York State Commission of Housing and Regional Planning to Governor Alfred E. Smith*, Albany, J. B. Lyon Co., May 7, 1926, reprinted in *Planning the Fourth Migration: The Neglected Vision of the Regional Planning Association of America*, ed. Carl Sussman (Cambridge, MA: MIT Press, 1976), 179.

9. Lewis Mumford, "The Fourth Migration," *Survey Graphic* (May 7, 1925); reprinted in Sussman, *Planning the Fourth Migration*, 55–64.

10. Mumford, "The Fourth Migration," 62.

11. Jonathan Sterne, "Transportation and Communication: Together as You've Always Wanted Them," in *Thinking with James Carey: Essays on Communications, Transportation, History*, ed. Jeremy Packer and Craig Robertson (New York: Peter Lang, 2006), 117–135.

12. Mumford, "The Fourth Migration," 63.

13. Lewis Mumford, "Regional Planning," July 8, 1931, roundtable on regionalism, Institute of Public Affairs, University of Virginia, Avery Library, Columbia University; reprinted in Sussman, *Planning the Fourth Migration*, 207.

14. The first version of the book, *To-Morrow: A Peaceful Path to Real Reform* was published in 1898. Reprinted in 1902, its title would become *Garden Cities of To-Morrow*: Ebenezer Howard, *Garden Cities of To-Morrow*, ed. F. J. Osborn (Cambridge, MA: MIT Press, [1946] 1965).

15. Lewis Mumford, "The Garden City Idea and Modern Planning," in Howard, *Garden Cities of To-Morrow*, 29.

16. Thomas P. Hughes, *Networks of Power: Electrification in Western Society 1880–1930* (Baltimore, MD: Johns Hopkins University Press, 1983).

17. Hughes, *Networks of Power*, 364.

18. One can compare this evolution to the process of concretization of the technical object that Georges Simondon describes in his work *Du mode d'existence des objets techniques*.

19. U.S. Department of Commerce, Bureau of the Census, *Historical Statistics of the United States: Colonial Times to 1957* (Washington, DC: GPO, 1960), 432–433 and 589. Citation in Hughes, *Networks of Power*, 365.

20. Hopkins, *Telephone Lines and Their Properties*, 111–112.

21. Ogle, *Long Distance Please*, 27.

22. Ogle, 37–38.

23. George W. Coy, Switch for Telephone Exchange, no. 224,653, patented February 17, 1880. Filed on May 20, 1879, this patent seems to be for the improvement of a previous version.

24. Horace Coon, *American Tel & Tel: The Story of a Great Monopoly* (Freeport, NY: Books of Libraries Press, 1971).

25. David E. Nye, *Electrifying America: Social Meanings of a New Technology, 1880–1940* (Cambridge, MA: MIT Press, 1992).

26. On the emergence of capitalist corporations in the United States, see David F. Noble, *America by Design: Science, Technology, and the Rise of Corporate Capitalism* (New York: Knopf, 1977).

27. Lewis Mumford, "Regional Planning," 204.

28. *Report of the New York State Commission of Housing and Regional Planning to Governor Alfred E. Smith*, Albany, NY: J. B. Lyon Co., May 7, 1926, reprinted in Sussman, *Planning the Fourth Migration*, 145–194.

29. Francesco Dal Co, "From Parks to the Region," in *The American City*, ed. Giorgio Ciucci, Francesco Dal Co, Mario Manieri-Elia, and Manfredo Tafuri (Cambridge, MA: MIT Press, [1973] 1983), 250.

30. In Sussman, *Planning the Fourth Migration*, 143.

31. Wright, *The Living City*.

32. On Jefferson, see Catherine Maumi, *Thomas Jefferson et le projet du Nouveau Monde* (Paris: Éditions de la Villette, 2007).

33. Giorgio Ciucci, "The City in Agrarian Ideology and Frank Lloyd Wright: Origins and Development of Broadacre," in Ciucci et al., *The American City*, 301.

34. The "Usonian" is thus the single-story house imagined for the middle-class inhabitants of "Usonia." Wright used the term *Usonia* to underscore a mythical dimension to the geographical territory of the United States of North America and its occupants. According to Catherine Maumi, Wright employed the word during his stay in Europe in the years after 1910. It was not until 1925, however, that his first explication of the term appeared in "In the Cause of Architecture: The Third Dimension," in *Frank Lloyd Wright: Collected Writings, Volume 2*, ed. Bruce B. Pfeiffer (New York: Rizzoli, 1992), 214. Maumi, *Usonia ou le mythe de la ville-nature américaine*, 5–6.

35. Catherine Maumi, "Lire entre les lignes," *Frank Lloyd Wright: Broadacre City, la nouvelle frontière* (Paris: Éditions de la Villette, 2015), 57.

36. Frank Lloyd Wright, *Broadacre City. A New Community Plan*, in *America Builds: Source*

Documents in American Architecture and Planning, ed. Leland M. Roth (New York: Harper & Row, [1935] 1983), 484.

37. Frank Lloyd Wright, "The City," in *Modern Architecture, Being the Kahn Lectures for 1930* (Princeton: Princeton University Press, 1931), later published in *Frank Lloyd Wright: Collected Writings, Volume 2*, ed. Bruce B. Pfeiffer (New York: Rizzoli, 1992), 69–71.

38. Paul Virilio, *La vitesse de libération* (Paris: Galilée, 1995), 11; and Céline Flécheux, *L'horizon* (Paris: Klincksieck, 2014), 75.

39. Bernhard Siegert, "(Not) in Place: The Grid, or, Cultural Techniques of Ruling Spaces," *Cultural Techniques: Grids, Filters, Doors, and Other Articulations of the Real* (New York: Fordham University Press, 2015), 98.

40. Rosalind Krauss, "Grids," *October* 9, no. 102 (1979): 50–64.

41. Krauss, "Grids," 52.

42. Georges Didi-Huberman, *Ce que nous voyons, ce qui nous regarde* (Paris: Les Éditions de Minuit, 1992).

43. Lewis Mumford, "The Sky Line: Mr. Wright's—Downtown Dignity," *New Yorker II* (April 27, 1935): 64, in *Frank Lloyd Wright and Lewis Mumford, Thirty Years of Correspondence*, ed. Bruce Brooks Pfeiffer and Robert Wojtowicz (New York: Princeton Architectural Press, 2001), 18–19.

44. Mumford, "The Sky Line."

45. Lewis Mumford, *The Story of Utopias* (New York: Boni and Liveright, 1922), 212–213.

46. Emily Bills, "Connecting Lines: L.A.'s Telephone History and the Binding of the Region," *Southern California Quarterly* 91, no. 1 (2009): 27–67.

47. Cf. Bills, "Connecting Lines," 27, note 1. See also Los Angeles Board of Public Utilities, Fifth Annual Report, 1914, Los Angeles City Archives, diagram 68, fig.13.

48. Specifically, Bills studied the case of the town of Pomona, which developed rapidly in 1880 and became one of the major citrus growing centers. The town's residents, organized into a cooperative, eventually created a local telephone company to establish a network with neighboring towns and meet the needs of their emerging economy: the Pomona Valley Telephone and Telegraph Union, intended to compete with the main company in the area, the Sunset Company, which was more interested in long-distance connections. See Bills, "Connecting Lines," 52–64.

49. Easterling, *Extrastatecraft: The Power of Infrastructure Space.*

Chapter 4

1. Letter from Alexander Graham Bell "to the capitalists of the Electric Telephone Company," in Mackenzie, *Alexander Graham Bell*, 203.

2. Mackenzie, *Alexander Graham Bell*, 203.

3. Mackenzie, 203.

4. Mackenzie, 205–206.

5. Marx, *The Machine in the Garden.*

6. Guilbert, "Très Haute Tension sur l'Architecture," 27–29.

7. Robert L. Fishman, *Urban Utopias in the Twentieth Century: Ebenezer Howard, Frank Lloyd Wright, and Le Corbusier* (Cambridge, MA: MIT Press, 1982), 3. If Lewis Mumford, rather than Ebenezer Howard, will be discussed in this chapter, this choice was dictated by the fact that while fundamentally linked to the thinking of the designer of the garden cities, the regionalist theories of Mumford and his colleagues develop more explicitly around telecommunications systems.

8. Ebenezer Howard, "Spiritual Influences Towards Social Progress," *Light* (April 30, 1910). See Robert Beevers, *The Garden City Utopia: A Critical Biography of Ebenezer Howard* (Basingstoke: Macmillan, 1988).

9. Stephen V. Ward, "Ebenezer Howard: His Life and Times," in *From Garden City to Green City: The Legacy of Ebenezer Howard*, ed. Kermit Personet and David Schuyler (Baltimore / London: The John Hopkins University Press, 2002), 19.

10. See Mumford, "The Fourth Migration," 55–64 and 199–208.

11. On the rhetoric of telecommunications technologies and its recurring myths, see Marvin, *When Old Technologies Were New*, 191–235, and Vincent Mosco, "When Old Myth Were New," in *The Digital Sublime: Myth, Power, and Cyberspace* (Cambridge, MA: MIT Press, 2004), 117–140.

12. The way in which the new media extended the space of domestic intimacy into mobility has been theorized by Raymond Williams through the concept of "mobile privatization." See Raymond Williams, *Television: Technology and Cultural Form* (London: Fontana, 1974).

13. Mumford, "The Fourth Migration," 63.

14. Susan Douglas, "Amateur Operators and American Broadcasting. Shaping the Future of Radio," in *Imagining Tomorrow: History, Technology, and the American Culture*, ed. Joseph J. Corn (Cambridge, MA: MIT Press, 1986), 35–57.

15. In 1909, Gernsback opened the first radio store. The following year, he formed the first amateur radio society and published the *Wireless Bluebook* and the first book on radio transmission, *The Wireless Telephone*. See Sam Moskowitz, *Hugo Gernsback: Father of Science-Fiction* (New York: Criterion Linotyping & Printing Co. 1959), 20.

16. Hugo Gernsback, *Radiocracy*, Christmas and New Year's brochure, 1944, 3, George H. Clark Radioana Collection, Archives Center, National Museum of American History, Smithsonian Institution.

17. Gilles Delalex, "J. G. Ballard, urbaniste du future," in *L'Infraville: futurs des infrastructures*, ed. Dominique Rouillard (Paris: Archibooks / Sautereau Éditeur, 2011), 84–93.

18. Delalex, "J. G. Ballard, urbaniste du future," 93.

19. Dominique Rouillard, "Introduction," in *Imaginaires d'infrastructure*, ed. Dominique Rouillard (Paris: L'Harmattan, 2009), 8.

20. Dominique Rouillard, "Le futur au travail," in *Imaginaires d'infrastructure*, 55–65.

21. In particular, Mumford sent the journal inventions like the "improved electrolytic detector," *Modern Electrics* (April 1912): 40; "a portable receiving outfit" (April 1911): 29–30; and the "ultimate crystal set" (June 1911): 179. See Arthur P. Molella, "Mumford in Historiographical Context," in *Lewis Mumford: Public Intellectual*, ed. Thomas P. and Agatha C. Hughes (New York / Oxford: Oxford University Press 1990), 21–42.

22. *Report of the New York State Commission of Housing and Regional Planning*, 145–194. Members of the RPAA included Clarence S. Stein, president of the New York State Commission, while Mumford and Benton MacKaye directed the study, and Henry Wright contributed ideas as well as illustrations. Stein was also the chief architect for the towns of Radburn and Sunnyside, working with Wright as well.

23. Robert Wojtowicz, *Lewis Mumford and American Modernism. Eutopian Theories for Architecture and Urban Planning* (Cambridge / New York: Cambridge University Press, 1996), 1.

24. Mumford, *The Story of Utopias*, 229–230.

25. Mumford, *The Story of Utopias*, 168–169. On the development of eutopian thought in totalitarian planning models, with the aid of modern technologies, see Robert H. Kargon and Arthur P. Molella, *Invented Edens, Techno-Cities of the Twentieth Century* (Cambridge, MA: MIT Press, 2008).

26. According to Wojtowicz, "in his writings, he (Mumford) repeatedly attempted to guide the reader not toward an impossible utopia, but rather toward 'eutopia,' the good place, brought down from the clouds and planted firmly on the ground." In Wojtowicz, *Lewis Mumford and American Modernism*, 1.

27. Geddes, *Cities in Evolution*, 73–74.

28. Geddes, 73.

29. Howard P. Segal, "Mumford's Alternative to the Megamachine: Critical Utopianism, Regionalism, and Decentralization," in *Lewis Mumford: Public Intellectual*, ed. Thomas P. and Agatha C. Hughes (New York / Oxford: Oxford University Press, 1990), 100–109. Segal refers to a "critical utopianism" to describe Mumford's position. "Mumford . . . believes not in the possibility of genuine utopias in the sense of literally perfect societies but rather in the utility of utopias. He allows that, whether as written blueprints or as actual communities, utopias can change existing societies. I call this alternative 'critical utopianism,'" 101.

30. See Frank Lloyd Wright, *The Disappearing City* (New York: W. F. Payson, 1932); Wright, *When Democracy Builds* (Chicago: University of Chicago Press, 1945); Wright, *The Living City*.

31. "Birds sing for him, grass grows green for him, rain falls for him on his growing crops while the wheels of standardization and money-making invention no longer turn dead against him. All turn now for his majesty, the American citizen!" in Wright, *The Living City*, 153.

32. Wright, *Broadacre City: A New Community Plan*, 487.

33. For the specific context of American utopianism and its connections with technological progress at the end of the nineteenth century and the beginning of the twentieth, see Howard P. Segal, *Technological Utopianism in American Culture* (Syracuse: Syracuse University Press, 2005).

34. Le Corbusier, *La ville radieuse* (Paris: Éditions Vincent Fréal & Cie, [1933] 1964). Le Corbusier presented "La ville contemporaine de trois millions d'habitants" at the Salon d'Automne of 1922. It was his first systematic attempt at formalizing his ideas on urban planning.

35. Le Corbusier, *La ville radieuse*, 38.

36. Le Corbusier, *Le Poème de l'angle droit* (Paris: Tériade, 1955), 150; English translation by Kenneth Hylton (Paris: Ed. Connivences, 1989), n.p.

37. Le Corbusier, *La ville radieuse*.

38. Albert Robida, *Le vingtième siècle: La vie électrique* (Paris: Librairie illustrée, 1893).

39. Fishman, *Urban Utopias*, 236, note 3.

40. For Le Corbusier's involvement in French syndicalism and the St. Simonian tradition, see Kenneth Frampton, "The City of Dialectic," *Architectural Design* 39 (October 1969): 541–543 and 545–546; republished in *Le Corbusier in Perspective*, ed. Peter Serenyi (Englewood Cliffs, NJ: Prentice-Hall Inc., 1975), 139–143.

41. Marc Duboiset and Yaron Pesztat, eds., *Architecture du telephone* (Brussels: AAM Editions, 2019); and Carlo Olmo and Alessandro De Magistris, eds., *Iakov Tchernikov* (Paris: Éditions d'art Somogy, 1995).

42. Fanny Lopez, "La perte du lieu: les infrastructures du capitalocène," in *Politique des infrastructures*, ed. Dominique Rouillard (Geneva: Mētis-Presses, 2018), 81–93.

43. Graham and Marvin, *Splintering Urbanism*.

Chapter 5

1. John Durham Peters, *The Marvelous Clouds: Toward a Philosophy of Elemental Media* (Chicago / London: The University of Chicago Press, 2015), 36.

2. Bertolt Brecht, "Radio as a Means of Communication: A Talk on the Function of Radio," *Screen* 20, no. 3–4 ([1932] Winter 1979): 24–28.

3. Alfredo Thiermann Riesco, "Radio Activities: The Architecture of Broadcasting in Cold War Berlin," PhD diss., Swiss Federal Institute of Technology Zurich, 2020 (Philip Ursprung and Jean-Louis Cohen, thesis advisors).

4. Nikolai Punin, "The Monument to the Third International," in *Art in Theory 1900–1990: An Anthology of Changing Ideas*, ed. Charles Harrison and Paul Wood (Malden, MA: Blackwell, [1920] 1992), 311–315.

5. Mark Wigley, *Buckminster Fuller Inc., Architecture in the Age of Radio* (Zurich: Lars Müller, 2015), 30–37.

6. Wigley, *Buckminster Fuller Inc.*, 35.

7. Wigley, 46.

8. Wigley, 26.

9. Thiermann, "Radio Activities: The Architecture of Broadcasting in Cold War Berlin," 26–27.

10. Carlotta Darò, "De l'Imperial Wireless Chain au Regional African Satellite Communications Organisation: l'Afrique sans fil," in *Afriques: architectures, infrastructures et territoires en devenir*, ed. Dominique Rouillard (Paris: ENSBA, 2015), 97–104.

11. Marc Raboy, *Marconi: The Man Who Networked the World* (Oxford: Oxford University Press, 2016).

12. Harold A. Innis, *Empire and Communications* (Oxford: Oxford University Press, 1950).

13. Friedrich Kittler, *Gramophone, Film, Typewriter* (Stanford: Stanford University Press, 1999).

14. Based on Marxian philosophy, the "technological determinism" theory assumes that technology guides the social structure and cultural values of an era. The term was originally used by the American economist Thorstein Veblen.

15. Michael Darroch, "Bridging Urban and Media Studies: Jaqueline Tyrwhitt and the Exploration Group 1951–1957," *Canadian Journal of Communication* 33, no. 2 (2008): 147–169.

16. Carlton D. Williams, "Acoustic Space," *Explorations: Studies in Culture and Communication* 4 (February 1955): 15–20. See Richard Carvell, *McLuhan in Space: A Cultural Geography* (Toronto: University of Toronto Press, 2002), 16–23.

17. Marshall H. McLuhan, *Understanding Media: The Extensions of Man* (Cambridge, MA: MIT Press, [1964] 1994), 283.

18. Hugo Gernsback, *Radio for All* (Philadelphia / London: J. B. Lippincott Company, 1922), reprint by General Books, 2010, 3–5.

Chapter 6

1. John B. Jackson, "Roads Belong in the Landscape," excerpts from *A Sense of Place, a Sense of Time* (New Haven: Yale University Press, 1994), published in *Landscape in Sight: Looking at America*, ed. Helen Lefkowitz Horowitz (New Haven: Yale University Press, 1997), 249–254.

2. Gérard Wajcman, *Fenêtre: Chroniques du regard et de l'intime* (Lagrasse: Verdier, 2004), 439–441; and Georges Teyssot, "Windows and Screens: A Topology of the Intimate and the Extimate," *Log*, no. 18 (2010), 75–88.

3. Wajcman, *Fenêtre*, 439.

4. Georges Duby, "Ouverture," in *Histoire de la vie privée*, vol. 2, *De l'Europe féodale à la Renaissance*, ed. Philippe Ariès and Georges Duby (Paris: Seuil, 1985), 19–24.

5. Duby, "Ouverture," 22–23.

6. Keasbey, *The Law of Electric Wires in Streets and Highways*, 1.

7. Keasbey, 110–111.

8. Julia Building Association v. Bell Telephone Company, Saint Louis, Missouri, 1885; Keasbey, *The Law of Electric Wires in Streets and Highways*, 114.

9. Hewett v. Western Union Telegraph Co, Washington D.C., 1886; Keasbey, *The Law of Electric Wires in Streets and Highways*, 115.

10. Roake v. American Telegraph & Telephone, New Jersey, 1886; Keasbey, *The Law of Electric Wires in Streets and Highways*, 117.

11. Keasbey, *The Law of Electric Wires in Streets and Highways*, 2.

12. Keasbey, 2.

13. François Béguin, "Savoirs de la ville et de la maison au début du 19e siècle," in *Politiques de l'habitat, 1800–1850*, ed. Michel Foucault (Paris: Corda, 1977), 211–324.

14. Béguin, "Savoirs de la ville et de la maison au début du 19e siècle," 247–251.

15. Béguin, 317–324.

16. Stephen Kern, *The Culture of Time and Space, 1880–1918* (Cambridge, MA: Harvard University Press, 1983), 187; Christoph Asendorf, "Telefon: 'Piercing the Shell of Privacy,'" *Arch+*, no.191/192 (March 2009): 118–123.

17. Beatriz Colomina, *Privacy and Publicity: Modern Architecture as Mass Media* (Cambridge, MA: MIT Press, 1994), 7–11.

18. Evelyne Patlagean, "Byzance Xe-XIe siècle," in *Histoire de la vie privée*, vol. 1, *De l'Empire romain à l'an mille*, ed. Philippe Ariès and Georges Duby (Paris: Seuil, 1985), 547–548; and Spiro Kostof, *A History of Architecture: Settings and Rituals* (New York: Oxford University Press, 1985), 361. See also Terence Riley, *The Un-Private House* (New York: Museum of Modern Art, 1999), 10.

19. Riley, *The Un-Private House*, 11; and Roger Chartier, "La pratique de l'écrit," in *Histoire de la vie privée*, vol. 3, *De la Renaissance aux Lumières*, ed. Philippe Ariès and Georges Duby (Paris: Seuil, 1986), 165–167.

20. Kazys Varnelis and Anne Friedberg, "Place: The Networking of Public Space," in *Networked Publics*, ed. Kazys Varnelis (Cambridge, MA: MIT Press, 2008), 18. See Jane Jacobs, *The Death and Life of Great American Cities* (New York: Random House, 1961).

21. Peter Sloterdijk, *Ecumes: Spheres III*, trans. Olivier Mannoni (Paris: Hachette Littératures, 2005), 443.

22. See Caroline A. Jones, ed., *Sensorium: Embodied Experiences, Technology, and Contemporary Art* (Cambridge, MA: MIT Press, 2006), 5–49.

23. Of course, other forms of mediation such as letters or telegrams had anticipated the possibility of a remote communication, requiring, however, a time of elaboration for the interpretation of the message.

Chapter 7

1. Jonathan Sterne, *The Audible Past* (Durham, NC: Duke University Press, 2003), 1–2.

2. David Edgerton, *The Shock of the Old* (London: Profile Books, 2006).

3. Ithiel de Sola Pool, Craig Decker, Stephen Dizard, Kay Israel, Pamela Rubin, and Barry Weinstein, "Foresight and Hindsight: The Case of the Telephone," in *The Social Impact of the Telephone*, 141.

4. American Telegraph and Telephone Company, *Planning for Home Telephone Conveniences: For the Use of Architects, Engineers, Builders, Owners* (1928), 7.

5. American Telegraph and Telephone Company, *Planning for Home Telephone Conveniences.*

6. On the definition and diffusion of the concept of acousmatic sound, see Pierre Schaeffer, *Traité des objets musicaux: Essais interdisciplinaires* (Paris: Éditions du Seuil, 1966) and Michel Chion, *Guide des objets sonores: Pierre Schaeffer et la recherche musicale* (Paris, Buchet-Chastel, 1983). For a critical reading of acousmatics, see Brian Kane, *Sound Unseen: Acousmatic Sound in Theory and Practice* (Oxford: Oxford University Press, 2014).

7. American Telegraph and Telephone Company, *Modern Telephone Service for the Home* (1929), unpaginated.

8. Jonathan Sterne, "Sound Out of Time / Modernity's Echo," in *Turning the Century: Essays in Media and Cultural Studies* ed. Carol A. Stabile (Boulder, CO: Westview Press, 2000), 9–30.

9. Adrian Forty, *Objects of Desire: Design & Society from Wedgwood to IBM* (New York: Pantheon Books, 1986). See esp. chapter 8, "Electricity—The Fuel of the Future," 182–206.

10. American Telegraph and Telephone Company, *Modern Telephone Service for the Home.*

11. Advertising brochure for Bell System (New Jersey Bell Telephone Company), AT&T, n.d.

12. Yve-Alain Bois, "Metamorphosis of Axonometry," *Daidalos* 1 (1981): 40–58; and Bruno Reichlin, "The Axonometric as a Project: A Study of Alberto Sartoris," *Lotus International* 22 (1979): 83–93. See Erwin Panofsky, *Perspective as Symbolic Form*, trans. Christopher S. Wood (New York: Zone Books, 1997).

13. Banham, *The Architecture of the Well-Tempered Environment*, 134–136.

14. For a reading of the usage of perspective in Western culture, see Massimo Scolari, *Il*

disegno obliquo: una storia dell'antiprospettiva (Venice: Marsilio Editori, 2005), especially the chapter "Disegnare in paralleli modo," 284–293.

15. Leon Battista Alberti, *De re aedificatoria*, 1485, book 2, chapter 1. For the English translation, see Leon Battista Alberti, *On the Art of Building in Ten Books*, trans. Joseph Rykwert, Neil Leach, and Robert Tavernor (Cambridge, MA: MIT Press, 1991).

16. See chapters on "Machinationes" and "La prospettiva soldatesca" in Scolari, *Il disegno obliquo*, 229–284. See also Jacques Guillerme, Hélène Vérin, and Stephen Sartarelli, "The Archeology of Section," *Perspecta* 25 (1989): 226–257.

17. Joël Sakarovitch, "Architecture et représentation: de la projection à la double projection," *In Extenso*, ed. Roger Laurent, no. 13 (1990): 63–101.

18. Sakarovitch, "Architecture et représentation," 90.

19. See Yve-Alain Bois, "Sergei M. Eisenstein: Montage and Architecture," *Assemblage* 10 (1989): 111–131.

20. Georges Teyssot, "A Topology of Thresholds," in *Home Culture* 2, no.1 (2005): 89–116.

21. Antoine Picon, "Du traité à la revue, l'image d'architecture au siècle de l'industrie," in *Usage de l'image au XIXe siècle*, ed. Stéphane Michaud, Jean-Yves Mollier, and Nicole Savy (Paris: Créaphis, 1992), 153–165.

22. Bruno Latour, "How to Be Iconophilic in Art, Science, and Religion?," in *Picturing Science, Producing Art*, ed. Carrie A. Jones and Peter Galison (London: Routledge, 1998), 418–440.

23. Pierre Caye, *Durer: Éléments pour la transformation du système productif* (Paris: Les Belles Lettres, 2020), 331–332.

24. Philip Johnson and Alfred H. Barr Jr., *Machine Art* (New York: Museum of Modern Art, 1934), 3.

25. Lewis Mumford, *Technics and Civilization* (New York: Harcourt, Brace & World, [1934] 1963), 12.

Chapter 8

1. Victor Hugo, *Choses vues: Œuvres complètes*, book 16 (Paris: Le Club français du livre, 1970), 911.

2. For a complete and documented history of the théâtrophone, including the relationship of this invention to the theater, see the publications of Melissa van Drie: "L'espace scénique du Théâtrophone (1881–1930) et la figure nouvelle du spectateur-auditeur," *Cahiers Création Arts et Patrimoines: Modèles et modalités de la transmission culturelle* 2 (2015): 41–68; "Devenir auditeur: une nouvelle conscience du son et de soi à travers l'expérience du théâtrophone (1881–1936)," in *Musique et enregistrement*, ed. Pierre-Henry Frangne and Hervé Lacombe (Rennes: PUR, 2014), 55–76.

3. Melissa van Drie "Hearing through the Théâtrophone: Sonically Constructed Spaces and Embodied Listening in Late Nineteenth-Century French Theatre," *Sound Effects: An Interdisciplinary Journal of Sound and Sound Experience* 5, no.1 (2015): 73–90.

4. Matei Calinescu, *Five Faces of Modernity: Modernism, Avant-garde, Decadence, Kitsch, Postmodernism* (Durham, NC: Duke University Press, 1987), 44.

5. Walter Benjamin, "Le Paris du second Empire chez Baudelaire," in *Charles Baudelaire: Un poète lyrique à l'apogée du capitalisme*, trans. Jean Lacoste (Paris: Payot, [1938] 1982), 100.

6. Julien Lefèvre, *L'électricité au théâtre* (Paris: A. Grelot, 1894), 322–337.

7. Jacques Attali and Yves Stourdzé, "The Birth of the Economic Crisis: The Slow Death of Monologue in French Society," in Pool, *The Social Impact of the Telephone*, 97–111.

8. Claude Chappe, *Lettres sur le nouveau télégraphe (de Bréguet et Bétancourt)* (Paris: 1798).

9. Edward Bellamy, *Looking Backward: 2000–1887* (New York: Random House, [1888] 1917), 222–223.

10. Lefèvre, *L'électricité au théâtre*, 322–337.

11. On acousmatic sound, see chapter 7, note 6.

12. Kane, *Sound Unseen*.

13. Gooday, *Domesticating Electricity*.

14. Michèle Martin, "Gender and Early Telephone Culture," in *The Sound Studies Reader*, ed. Jonathan Sterne (New York: Routledge, 2012), 336–350.

15. Martin, "Gender and Early Telephone Culture," 339.

16. Danièle Laster, "Splendeurs et misères du théâtrophone," *Romantisme*, no. 41 (1983): 74–78.

17. Jonathan Crary, *Techniques of the Observer: On Vision and Modernity in the Nineteenth Century* (Cambridge, MA: MIT Press, 1990).

18. Eugène Testavin, "L'organisation actuelle du théâtrophone," *Annales des Postes, télégraphes et téléphones* 19, no 1 (1930): 8.

19. Benjamin, "Le Paris du second Empire chez Baudelaire," 179.

20. Benjamin, 207.

21. Marcel Proust, *À un ami, correspondance de Marcel Proust*

avec Georges de Lauris (Paris: Amiot-Dumont, 1948), 243.

22. Marcel Proust, *Lettres à Reynaldo Hahn* (Paris: Gallimard, 1956), 199.

23. Laster, "Splendeurs et misères du théâtrophone," 78.

24. Testavin, *L'organisation actuelle du théâtrophone*, 4.

25. Testavin, 24.

26. Sterne, "Sound Out of Time," 15.

27. Van Drie, *Hearing through the Théâtrophone*, 65.

28. Carlotta Darò, *Avant-gardes sonores en architecture* (Dijon: Les presses du réel, 2013), 120–134.

Chapter 9

1. Benjamin Mueller, "On Crowded London Streets, Councils Fight a Flood of Phone Boxes," *New York Times* (September 29, 2018), accessed January 10, 2023, https://www.nytimes.com/2018/09/29/world/europe/london-phone-boxes-inlink.html.

2. Palko Karasz, "The Red Phone Box, a British Icon, Stages a Comeback," *New York Times* (May 22, 2018), accessed January 10, 2023, https://www.nytimes.com/2018/05/22/world/europe/uk-red-phone-box.html.

3. Cécile Ducourtieux, "En 1881, la première cabine était comparée à un confessionnel," *Le Monde*, May 23 2009, accessed June 25 2023, https://www.lemonde.fr/economie/article/2009/05/23/en-1881-la-premiere-cabine-etait-comparee-a-un-confessionnal_1197072_3234.html.

4. Vittorio Magnago Lampugnani, *Frammenti urbani: I piccoli oggetti che raccontano le città* (Torino: Bollati Boringheri, 2021), 53.

5. Lampugnani, *Frammenti urbani*, 54.

6. William Gray, Coin-Controlled Apparatus for Telephones, no. 408,709, patented August 13, 1889.

7. Richard Busteed, Telephone Cabinet, no. 282,841, patented August 7, 1883.

8. David Massey, "Telephone Tribute," accessed January 10, 2023, http://telephonetribute.com/payphones.html.

9. Orange Historical collection, accessed January 10, 2023, https://histoire.orange.com/fr/nos_objets/cabine-telephonique-1900/.

10. Bruno Latour, "Le groom est en grève. Pour l'amour de Dieu fermez la porte," in *La clef de Berlin et autres leçons d'un amateur de sciences* (Paris: La Découverte, 1993), 56–76.

11. See chapter 8 in this book.

12. Brenda Maddox, "Women and the Switchboard," in Pool, *The Social Impact of the Telephone*, 262–280.

13. Fanny Carmagnat, "Une approche socio-technique de l'histoire du telephone public," *Réseaux* no. 115 (2002): 243–265.

14. On British phone booths, see Gavin Stamp, *Telephone Boxes* (London: Chatto & Windus, 1989), and Neil Johannessen, *Telephone Boxes* (Princes Risborough: Shire, 2010).

15. The British telephone booths were first auctioned in the mid-1980s, and again in 2012, by the British telecom company BT. Hayat Gazzane, "Des cabines téléphoniques britanniques en vente," *Le Figaro*, April 29, 2012, accessed June 25, 2022, https://www.lefigaro.fr/societes/2012/04/28/20005-20120428ARTFIG00320-des-cabines-telephoniques-britanniques-en-vente.php.

16. Claude Prelorenzo, "Le mythe cinématographique des infrastructures," in Rouillard, *Imaginaires d'infrastructure*, 15–27.

17. Renée Reizman, "What Killed the Pay Phone," *The Atlantic* (February 2, 2017), accessed June 25 2022, https://www.theatlantic.com/technology/archive/2017/02/object-lesson-phone-booth/515385/.

18. Reizman, "What Killed the Pay Phone."

Chapter 10

1. László Moholy Nagy, *The New Vision and Abstract of an Artist* (Mineola, NY: Dover Publication, [1947] 2005), 223.

2. Moholy Nagy, *The New Vision and Abstract of an Artist*, 223.

3. In her book from 1972, *Margin Notes: Documentary Absurdities*, Lucia Moholy-Nagy claimed that her husband had actually traveled to a sign fabricator to order his enamel paintings, but that when he returned, he exclaimed, "I could have done this over the telephone." Lucia Moholy-Nagy, *Margin Notes: Documentary Absurdities* (Krefeld: Scherpe Verlag, 1972).

4. Louis Kaplan, "The Telephone Paintings: Hanging Up Moholy," in *Leonardo* 26, no. 2 (April 1993): 165–168.

5. Although it had initially been scheduled for the spring of 1968, the exhibition was canceled due to technical problems. It was ultimately held at the Museum of Contemporary Art of Chicago.

6. "UbuWeb: Sound," accessed March 26, 2020, ubu.com/sound/art_by_telephone.html.

7. Jan van der Marck in *Art by Telephone* (Chicago: Museum of

Contemporary Art, 1969). The exhibition catalogue was republished in 2008 by David Platzker as *Art by Telephone* (New York: Specific Objects, 2008).

8. With the active participation of students from the Ecole Supérieure des Beaux-Arts of Angers and Bordeaux, the Université Paris I Panthéon Sorbonne, along with students of Peter Tracey Connor and John Miller from Barnard College, Columbia University in New York. The exhibition was on view at the Cneai de Chatou, the CAPC in Bordeaux, at the Emily Harvey Foundation in New York, and the SFAI San Francisco, as well as in the exhibition halls of the Ecole Supérieure des Beaux-Arts of Angers.

9. McLuhan, *Understanding Media*, xi.

10. With John Cage, Lucinda Childs, Öyvind Fahlström, Alex Hay, Deborah Hay, Steve Paxton, Yvonne Rainer, Robert Rauschenberg, David Tudor and Robert Whitman.

11. Billy Klüver, 1988, text from the jacket cover of *Variations VII* by John Cage, E.A.T. and ARTPIX, 2008.

12. Hopkins, *Telephone Lines and Their Properties*, 113.

13. Maddox, "Women and the Switchboard," 264. On the medical origins and the relations between otology and the history of sound technologies, see Sterne, *The Audible Past*.

14. On this material approach, see Adrien Forty, *Concrete and Culture: A Material History* (London: Reaktion Books, 2012).

15. In his book *Keywords*, Raymond Williams defines "nature" and "culture," two of the most difficult and complex words in the English language, and in other European idioms, because of their semantic evolution over time. One of the most widespread uses, established in the eighteenth century, sees these two terms in categorical opposition, with nature representing a form of innocence that precedes the works of human creation, and culture conversely as the result of the development of human civilization. Williams, *Keywords*, 87 and 219.

16. We return here to the complexity of the thing, which, according to Heidegger, allows us to distinguish the singular nature of each object in its space and time of existence. Heidegger, "The Thing."

Aitchison, Mathew. "The Boyd Ultimatum." AA *Files* 66 (2013): 59–67.

American Telegraph and Telephone Company. *Modern Telephone Service for the Home*, 1929.

American Telegraph and Telephone Company. *Planning for Home Telephone Conveniences: For the Use of Architects, Engineers, Builders, Owners*, 1928.

Asendorf, Christoph. "Telefon: 'Piercing the Shell of Privacy.'" *Arch+*, no. 191/192 (March 2009): 118–123.

Attali, Jacques, and Yves Stourdzé. "The Birth of the Economic Crisis: The Slow Death of Monologue in French Society." In *The Social Impact of the Telephone*, edited by Ithiel de Sola Pool, 97–111. Cambridge, MA: MIT Press, 1977.

Bakke, Gretchen. *The Grid: The Fraying Wires Between Americans and Our Energy Future*. New York / London: Bloomsbury, 2016.

Bancaud, Florence. "L'esthétique du laid, de Hegel à Rosenkranz. Une 'esthétique de la résistance' ou de la résignation aux 'arts qui ne sont plus beaux'?" *Études Germaniques* 256, no. 4 (2009): 899–917.

Banham, Reyner. *The Architecture of the Well-Tempered Environment*. London: Architectural Press, 1969.

Barney, Darin. "Pipelines." In *Fueling Culture: Politics, History, Energy*, edited by Imre Szeman, Jennifer Wenzelt, and Patsy Yaeger, 267–270. New York: Fordham University Press, 2017.

Baudez, Gildas, and François Béguin. *Critique du paysage de l'énergie*. Paris: Roux-Bauer and Associates, 1980.

Baudin, Antoine. "Les constructeurs de Fernand Léger: De la place Smolensk à la vallée de Chevreuse, ou quelques tribulations d'un chantier pictural." *Matières*, no. 2 (1998): 68–75.

Baudrillard, Jean. *Le système des objets*. Paris: Gallimard, 1968. Translated by James Benedict as *The System of Objects* (London / New York: Verso Books, 2020).

Beevers, Robert. *The Garden City Utopia: A Critical Biography of Ebenezer Howard*. Basingstoke: Macmillan, 1988.

Béguin, François. "La quincaillerie du nouveau monde." *Le genre humain*, no. 34 (1999): 149–162.

Béguin, François. "Savoirs de la ville et de la maison au début du 19e siècle." In *Politiques de l'habitat*, 1800–1850, edited by Michel Foucault, 211–324. Paris: Corda, 1977.

Bell, Alexander Graham. "To the capitalists of the Electric Telephone Company." In Catherine Dunlop Mackenzie. *Alexander Graham Bell: The Man Who Contracted Space*, 202-206.

Bellamy, Edward. *Looking Backward: 2000–1887*. New York: Random House, [1888] 1917.

Beltran, Alain, and Patrice A. Carré. *La fée et la servante: la société française face à l'électricité, XIXe–XXe siècles*. Paris: Éditions Belin, 1991.

Benjamin, Walter. "Le Paris du second Empire chez Baudelaire." In *Charles Baudelaire: Un poète lyrique à l'apogée du capitalisme*, translated by Jean Lacoste. Paris: Payot, [1938] 1982.

Bennett, Jane. *Vibrant Matter: A Political Ecology of Things*. Durham, NC: Duke University Press, 2010.

Berland, Jody. "Space at the Margins." In *North of Empire: Essays on the Cultural Technologies of Space*, 65–97. Durham, NC: Duke University Press, 2009.

Bills, Emily. "Connecting Lines: L.A.'s Telephone History and the Binding of the Region." *Southern California Quarterly* 91, no. 1 (2009): 27–67.

Bois, Yve-Alain. "Metamorphosis of Axonometry." *Daidalos* 1 (1981): 40–58.

Bois, Yve-Alain. "Sergei M. Eisenstein: Montage and Architecture." *Assemblage* 10 (1989): 110–131.

Boulet, Roger, ed. *Vistas: Artists on the Canadian Pacific Railway*. Calgary: Glenbow Museum, 2009.

Boyd, Robin. *The Australian Ugliness*. Melbourne: Cheshire, [1960] 1963.

Brecht, Bertolt. "Radio as a Means of Communication: A Talk on the Function of Radio." *Screen* 20, no. 3–4 ([1932] Winter 1979): 24–28.

Bristol, C. H. *The Western Union Telegraph Company: Rules for Construction and Repairs*. New York: The Western Union Telegraph Company, 1905.

Brooks, John. *Telephone: The First Hundred Years*. New York: Harper & Row Publishers, 1975.

Calinescu, Matei. *Five Faces of Modernity: Modernism, Avant-garde, Decadence, Kitsch, Post-modernism*. Durham, NC: Duke University Press, 1987.

Carmagnat, Fanny. "Une approche socio-technique de l'histoire du telephone public." *Réseaux* no. 115 (2002): 243–265.

Carvell, Richard. *McLuhan in Space: A Cultural Geography*. Toronto: University of Toronto Press, 2002.

Castells, Manuel. *End of Millennium*. Vol. 3, *The Information Age: Economy, Society and Culture*. Malden/Oxford: Blackwell, 1998.

Castells, Manuel. *The Power of Identity*. Vol. 2. *The Information Age: Economy, Society and Culture*. Malden/Oxford: Blackwell, 1997.

Castells, Manuel. *The Rise of the Network Society*. Vol. 1, *The Information Age: Economy, Society and Culture*. Malden/Oxford: Blackwell, 1996.

Caye, Pierre. *Durer: Éléments pour la transformation du système productif*. Paris: Les Belles Lettres, 2020.

Chabard, Pierre. "Cities in Evolution (1915): The Writing and the Readings of an Intempestive Classic." In *Manifestoes and Transformations in the Early Modernist City*, edited by Christian Hermansen, 149–162. London: Ashgate, 2010.
Chappe, Claude. *Lettres sur le nouveau télégraphe (de Bréguet et Bétancourt)*. Paris: 1798.
Charbon, Paul and Pierre Lajarrige. "Le téléphone ce jeune centenaire: le Théâtrophone." *Revue française des télécommunications* (October 21, 1976): 49–53.
Chartier, Roger. "La pratique de l'écrit." In *Histoire de la vie privée*, vol. 3, *De la Renaissance aux Lumières*, edited by Philippe Ariès and Georges Duby, 165–167. Paris: Seuil, 1986.
Chernikhov, Yakov. *Architectural Fantasies: 101 Compositions in Color, 101 Architectural Miniatures*. Leningrad: Mezhdunarodnaia kniga, 1933.
Chion, Michel. *Guide des objets sonores: Pierre Schaeffer et la recherche musicale*. Paris, Buchet-Chastel, 1983.
Colomina, Beatriz. *Privacy and Publicity: Modern Architecture as Mass Media*. Cambridge, MA: MIT Press, 1994.
Coon, Horace. *American Tel & Tel: The Story of a Great Monopoly*. Freeport, NY: Books of Libraries Press, 1971.
Crary, Jonathan. *Techniques of the Observer: On Vision and Modernity in the Nineteenth Century*. Cambridge, MA: MIT Press, 1990.
Dal Co, Francesco. "From Parks to the Region." In *The American City*, edited by Giorgio Ciucci, Francesco Dal Co, Mario Manieri-Elia, and Manfredo Tafuri, 143–292. Cambridge, MA: MIT Press, [1973] 1983.
Darò, Carlotta. *Avant-gardes sonores en architecture*. Dijon: Les presses du réel, 2013.
Darò, Carlotta. "De l'Imperial Wireless Chain au Regional African Satellite Communications Orgaisation: l'Afrique sans fil." In *Afriques: architectures, infrastructures et territoires en devenir*, edited by Dominique Rouillard, 97–104. Paris: ENSBA, 2015.
Darroch, Michael. "Bridging Urban and Media Studies: Jaqueline Tyrwhitt and the Exploration Group 1951–1957." *Canadian Journal of Communication* 33, no. 2 (2008): 147–169.
Delalex, Gilles. "J. G. Ballard, urbaniste du future." In *L'Infraville: futurs des infrastructures*, edited by Dominique Rouillard, 84–93. Paris: Archibooks / Sautereau Éditeur, 2011.
Didi-Huberman, Georges. *Ce que nous voyons, ce qui nous regarde*. Paris: Les Éditions de Minuit, 1992.
Douglas, Susan. "Amateur Operators and American Broadcasting. Shaping the Future of Radio." In *Imagining Tomorrow: History, Technology, and the American Culture*, edited by Joseph J. Corn, 35–57. Cambridge, MA: MIT Press, 1986.

Duboiset, Marc, and Yaron Pesztat, eds. *Architecture du telephone*. Brussels: AAM Editions, 2019.

Duby, Georges. "Ouverture." In *Histoire de la vie privée*, vol. 2, *De l'Europe féodale à la Renaissance*, edited by Philippe Ariès and Georges Duby, 19–24. Paris: Seuil, 1985.

Ducourtieux, Cécile. "En 1881, la première cabine était comparée à un confessionnel." *Le Monde* (May 23, 2009). Accessed June 25, 2023. https://www.lemonde.fr/economie/article/2009/05/23/en-1881-la-premiere-cabine-etait-comparee-a-un-confessionnal_1197072_3234.html.

Easterling, Keller. *Extrastatecraft: The Power of Infrastructure Space*. London: Verso, 2014.

Edgerton, David. *The Shock of the Old*. London: Profile Books, 2006.

Farhat, Georges. "L'urbain comme paysage infrastructurel: Le val de Bièvre." In *La métropole des infrastructures*, edited by Claude Prelorenzo and Dominique Rouillard, 268–281. Paris: Éditions Picard, 2009.

Fishman, Robert L. *Urban Utopias in the Twentieth Century: Ebenezer Howard, Frank Lloyd Wright, and Le Corbusier*. Cambridge, MA: MIT Press, 1982.

Flécheux, Céline. *L'horizon*. Paris: Klincksieck, 2014.

Forty, Adrian. *Concrete and Culture: A Material History*. London: Reaktion Books, 2012.

Forty, Adrian. *Objects of Desire: Design & Society from Wedgwood to IBM*. New York: Pantheon Books, 1986.

Frampton, Kenneth "The City of Dialectic." In *Le Corbusier in Perspective*, edited by Peter Serenyi, 139–143. Englewood Cliffs, NJ: Prentice-Hall Inc., 1975.

Frye, Northrop. "The Canadian Scene: Explorers and Observers." In *Canadian Landscape Painting, 1670–1930: The Artist and the Land*, edited by Robert Hamilton Hubbard, 1–29. Madison: University of Wisconsin Press, 1973.

Gazzane, Hayat. "Des cabines téléphoniques britanniques en vente," *Le Figaro* (April 29, 2012). Accessed June 25, 2022. https://www.lefigaro.fr/societes/2012/04/28/20005-20120428ARTFIG00320-des-cabines-telephoniques-britanniques-en-vente.php.

Geddes, Patrick. *Cities in Evolution: An Introduction to the Town Planning Movement and to the Study of Civics*. London: Williams & Norgate, 1915.

Geddes, Patrick. "Civics: As Applied Sociology. Part I." In *Sociological Papers 1904*, 103–118. London: Macmillan, 1905.

Gernsback, Hugo. *Radiocracy*. Christmas and New Year's brochure, 1944.

Gernsback, Hugo. *Radio for All*. Philadelphia / London: J. B. Lippincott Company, 1922.

Ghosn, Rania. "Territories of Oil: The Trans-Arabian Pipeline." In *The Arab City. Architecture and Representation*, edited by Amale Andraos and Nora Akawi, 164–175. New York: Columbia Books on Architecture and the City, 2016.

Giedion, Sigfried. *Mechanization Takes Command: A Contribution to Anonymous History*. New York: Oxford University Press, 1948.

Glickman, Susan. *The Picturesque and the Sublime: A Poetics of the Canadian Landscape*. Montreal / Kingston: McGill-Queen's University Press, 1998.

Gooday, Graeme. *Domesticating Electricity: Technology, Uncertainty and Gender, 1880–1914*. London: Pickering & Chatto, 2008.

Graf, Franz, and Giulia Marino, eds. *Les dispositifs du confort dans l'architecture du XXe siècle: connaissance et stratégie de sauvegarde*. Lausanne: Presses polytechniques et universitaires romandes, 2016.

Graham, Stephen, and Simon Marvin. *Splintering Urbanism: Networked Infrastructures, Technological Mobilities and the Urban Condition*. London: Routledge, 2001.

Grandville. *Un autre monde: Transformations, visions, incarnations, ascensions, locomotives, explorations, pérégrinations; explorations, excursions; stations, cosmogonies, fantasmagories, rêveries, folâtreries, facéties, lubies, métamorphoses, zoomorphoses, lithomorphoses, métempsycoses, apothéoses et autres choses*. Paris: H. Fournier, 1844.

Guilbert, Alexis. "Très Haute Tension sur l'Architecture: L'implication des architectes dans la conception de pylônes électriques." Master 2 thesis, ENSA Paris Malaquais, January 2013.

Guillerme, Jacques, Hélène Vérin, and Stephen Sartarelli. "The Archeology of Section." *Perspecta* 25 (1989): 226–257.

Harvey, David. *Spaces of Capital: Towards a Critical Geography*. Edinburgh: Edinburgh University Press, 2001.

Heidegger, Martin. "The Thing." In *Poetry, Language, Thought*, translated by Albert Hofstadter, 165–186. New York: Harper and Row, [1951] 1971.

Hopkins, William J. *Telephone Lines and Their Properties*. New York: Longmans, Green, and Co., 1899.

Howard, Ebenezer. *Garden Cities of To-Morrow*. Edited by F. J. Osborn. Cambridge, MA: MIT Press, [1946] 1965.

Hughes, Thomas P. *Networks of Power: Electrification in Western Society 1880–1930*. Baltimore, MD: Johns Hopkins University Press, 1983.

Hughes, Thomas P., and Agatha C. Hughes, eds. *Lewis Mumford: Public Intellectual*. New York / Oxford: Oxford University Press, 1990.

Hugo, Victor. *Choses vues: Œuvres complètes*, book 16, Paris: Le Club français du livre, 1970.

Ingold, Tim. *Une brève histoire des lignes*. Translated by Sophie Renaut. Brussels: Zones sensibles, [2007] 2013.

Innis, Harold A. *Empire and Communications*. Oxford: Oxford University Press, 1950.

Innis, Harold A. *Staples, Markets, and Cultural Change: Selected Essays*. Montreal / Kingston: McGill-Queen's University Press, 1995.

Jackson, John B. *Landscape in Sight: Looking at America*. Edited by Helen Lefkowitz Horowitz. New Haven: Yale University Press, 1997.

Jacobs, Jane. *The Death and Life of Great American Cities*. New York: Random House, 1961.

Jarrige, François. *Techno-critiques: Du refus des machines à la contestation des technosciences*. Paris: Éditions La Découverte, [2014] 2016.

Johannessen, Neil. *Telephone Boxes*. Princes Risborough: Shire, 2010.

John, Richard R. *Network Nation: Inventing American Telecommunications*. Cambridge / London: The Belknap Press of Harvard University Press, 2010.

Johnson, Philip, and Alfred H. Barr Jr. *Machine Art*. New York: Museum of Modern Art, 1934.

Jones, Caroline A. ed. *Sensorium: Embodied Experiences, Technology, and Contemporary Art*. Cambridge, MA: MIT Press, 2006.

Kane, Brian. *Sound Unseen: Acousmatic Sound in Theory and Practice*. Oxford: Oxford University Press, 2014.

Kaplan, Louis. "The Telephone Paintings: Hanging Up Moholy." *Leonardo* 26, no. 2 (April 1993): 165–168.

Karasz, Palko. "The Red Phone Box, a British Icon, Stages a Comeback." *New York Times* (May 22, 2018). Accessed January 10, 2023. https://www.nytimes.com/2018/05/22/world/europe/uk-red-phone-box.html.

Kargon, Robert H., and Arthur P. Molella. *Invented Edens, Techno-Cities of the Twentieth Century*. Cambridge, MA: MIT Press, 2008.

Keasbey, Edward Q. *The Law of Electric Wires in Streets and Highways*. Chicago: Callaghan and Company, 1900.

Kern, Stephen. *The Culture of Time and Space, 1880–1918*. Cambridge, MA: Harvard University Press, 1983.

Kittler, Friedrich. *Gramophone, Film, Typewriter*. Stanford: Stanford University Press, 1999.

Kostof, Spiro. *A History of Architecture: Settings and Rituals*. New York: Oxford University Press, 1985.

Krauss, Rosalind. "Grids." *October* 9, no. 102 (1979): 50–64.

Lampugnani, Vittorio Magnago. *Frammenti urbani: I piccoli oggetti che raccontano le città*. Torino: Bollati Boringheri, 2021.

Laster, Danièle. "Splendeurs et misères du théâtrophone." *Romantisme*, no. 41 (1983): 74–78.

Latour, Bruno. "How to Be Iconophilic in Art, Science, and Religion?" In *Picturing Science, Producing Art*, edited by Carrie A. Jones and Peter Galison, 418–444. London: Routledge, 1998.

Latour, Bruno. "Le groom est en grève. Pour l'amour de Dieu fermez la porte." In *La clef de Berlin et autres leçons d'un amateur de sciences*, 56–76. Paris: La Découverte, 1993.
Le Corbusier. *La ville radieuse*. Paris: Éditions Vincent Fréal & Cie, [1933] 1964.
Le Corbusier. *Le Poème de l'angle droit*. Paris: Tériade, 1955. Translated by Kenneth Hylton as *The Poem of the Right Angle*. Paris: Ed. Connivences, 1989.
Lefèvre, Julien. *L'électricité au théâtre*. Paris: A. Grelot, 1894.
Lepic, Andres. "Mies and Photomontage, 1910–38." In *Mies in Berlin*, edited by Terence Riley and Barry Bergdoll, 324–329. New York: Museum of Modern Art, 2001.
"L'Exposition d'électricité," *La Nature*, no. 434 (September 24, 1881): 17.
Lopez, Fanny. "La perte du lieu: les infrastructures du capitalocène." In *Politique des infrastructures*, edited by Dominique Rouillard, 81–93. Geneva: MētisPresses, 2018.
Mackenzie, Catherine Dunlop. *Alexander Graham Bell: The Man Who Contracted Space*. Boston: Houghton Mifflin Company, 1928.
Maddox, Brenda. "Women and the Switchboard." In *The Social Impact of the Telephone*, edited by Ithiel de Sola Pool, 262–280. Cambridge, MA: MIT Press, 1977.
Marchand, Roland. *Advertising the American Dream: Making Way for Modernity, 1920–1940*. Berkeley: University of California Press, 1985.
Mareschal, Georges. "Le Théâtrophone." *La Nature*, no. 995 (June 25, 1892): 55–58.
Marin, Martine. "Femmes et société: le travail ménager (1919–1939)." PhD diss., Université Paris-Diderot Paris 7, 1984.
Martin, Michèle. "Gender and Early Telephone Culture." In *The Sound Studies Reader*, edited by Jonathan Sterne, 336–350. New York: Routledge, 2012.
Marvin, Carolyn. *When Old Technologies Were New: Thinking About Electric Communication in the Late Nineteenth Century*. New York: Oxford University Press, 1988.
Marx, Leo. *The Machine in the Garden: Technology and the Pastoral Ideal in America*. London: Oxford University Press, 1964.
Massey, David. "Telephone Tribute." Accessed January 10, 2023. http://telephonetribute.com/payphones.html.
Mattern, Shannon. *Code and Clay, Data and Dirt: Five Thousand Years of Urban Media*. Minneapolis: University of Minnesota Press, 2017.
Maumi, Catherine. *Frank Lloyd Wright: Broadacre City, la nouvelle frontière*. Paris: Éditions de la Villette, 2015.
Maumi, Catherine. *Usonia ou le mythe de la ville-nature américaine*. Paris: éd. de la Villette, 2008.

Maumi, Catherine. *Thomas Jefferson et le projet du Nouveau Monde*. Paris: Éditions de la Villette, 2007.

McCormack, Derek P. *Atmospheric Things: On the Allure of Elemental Envelopment*. Durham, NC / London: Duke University Press, 2018.

McLuhan, Marshall H. *D'oeil à oreille*. Translated by Derrick de Kerckhove. Montreal: Éditions Hartubise HMH, 1977.

McLuhan, Marshall H. *Understanding Media: The Extensions of Man*. Cambridge, MA: MIT Press, [1964] 1994.

Moholy Nagy, László. *The New Vision and Abstract of an Artist*. Mineola: Dover Publication, [1947] 2005.

Moholy-Nagy, Lucia. *Margin Notes: Documentary Absurdities*. Krefeld: Scherpe Verlag, 1972.

Molella, Arthur P. "Mumford in Historiographical Context." In *Lewis Mumford: Public Intellectual*, edited by Thomas P. and Agatha C. Hughes, 21–42. New York / Oxford: Oxford University Press, 1990.

Mosco, Vincent. *The Digital Sublime: Myth, Power, and Cyberspace*. Cambridge, MA: MIT Press, 2004.

Moskowitz, Sam. *Hugo Gernsback. Father of Science-Fiction*. New York: Criterion Linotyping & Printing Co. 1959.

Mueller, Benjamin. "On Crowded London Streets, Councils Fight a Flood of Phone Boxes." *New York Times* (September 29, 2018). Accessed January 10, 2023. https://www.nytimes.com/2018/09/29/world/europe/london-phone-boxes-inlink.html.

Mumford, Lewis. "The Garden City Idea and Modern Planning" In Ebenezer Howard, *Garden Cities of To-Morrow*, edited by F. J. Osborn, 29–40. Cambridge, MA: MIT Press, [1946] 1965.

Mumford, Lewis. *Planning the Fourth Migration: The Neglected Vision of the Regional Planning Association of America*. Edited by Carl Sussman. Cambridge, MA: MIT Press, 1976.

Mumford, Lewis. *The Story of Utopias*. New York: Boni and Liveright, 1922.

Mumford, Lewis. *Technics and Civilization*. New York: Harcourt, Brace & World, [1934] 1963.

Noble, David F. *America by Design: Science, Technology, and the Rise of Corporate Capitalism*. New York: Knopf, 1977.

Novak, Barbara. *Nature and Culture: American Landscape and Painting*. London: Thames & Hudson, 1980.

Nye, David E. *American Technological Sublime*. Cambridge, MA: MIT Press, 1994.

Nye, David E. *Electrifying America: Social Meanings of a New Technology, 1880–1940*. Cambridge, MA: MIT Press, 1992.

Ogle, Edmond B. *Long Distance Please: The Story of the Trans Canada Telephone System*. Toronto: Collins, 1979.

Olmo, Carlo, and Alessandro De Magistris, eds. *Iakov Tchernikov*. Paris: Éditions d'art Somogy, 1995.

Patlagean, Evelyne. "Byzance Xe–XIe siècle." In *Histoire de la vie privée*, vol. 1, *De l'Empire romain à l'an mille*, edited by Philippe Ariès and Georges Duby, 547–548. Paris: Seuil, 1985.

Perrot, Michelle. "Histoire de la condition féminine et histoire de l'électricité." In *Électricité dans l'histoire: problèmes et méthodes*. Conference proceedings of the Association pour l'histoire de l'électricité en France, Paris, October 11–13, 1983.

Personet, Kermit, and David Schuyler, eds. *From Garden City to Green City: The Legacy of Ebenezer Howard*. Baltimore / London: The John Hopkins University Press, 2002.

Peters, John Durham. *The Marvelous Clouds: Toward a Philosophy of Elemental Media*. Chicago: University of Chicago Press, 2015.

Pfeiffer, Bruce Brooks, ed. *Frank Lloyd Wright: Collected Writings, Volume 2*. New York: Rizzoli, 1992.

Pfeiffer, Bruce Brooks, and Robert Wojtowicz, eds. *Frank Lloyd Wright and Lewis Mumford: Thirty Years of Correspondence*. New York: Princeton Architectural Press, 2001.

Picon, Antoine. "Du traité à la revue, l'image d'architecture au siècle de l'industrie." In *Usage de l'image au XIXe siècle*, edited by Stéphane Michaud, Jean-Yves Mollier and Nicole Savy, 153–165. Paris: Créaphis, 1992.

Ponte, Alessandra. "Journey to the North." In *The House of Light and Entropy*, 135–163. London: Architectural Association, 2014.

Pool, Ithiel de Sola, ed. *The Social Impact of the Telephone*. Cambridge, MA: MIT Press, 1977.

Pool, Ithiel de Sola, Craig Decker, Stephen Dizard, Kay Israel, Pamela Rubin, and Barry Weinstein. "Foresight and Hindsight: The Case of the Telephone." In *The Social Impact of the Telephone*, edited by Ithiel de Sola Pool, 127–157.

Prelorenzo, Claude. "Le mythe cinématographique des infrastructures." In *Imaginaires d'infrastructure*, edited by Dominique Rouillard, 15–27. Paris: L'Harmattan, 2009.

Proust, Marcel. *À un ami, correspondance de Marcel Proust avec Georges de Lauris*. Paris: Amiot-Dumont, 1948.

Proust, Marcel. *Lettres à Reynaldo Hahn*. Paris: Gallimard, 1956.

Punin, Nikolai. "The Monument to the Third International." In *Art in Theory 1900–1990: An Anthology of Changing Ideas*, edited by Charles Harrison and Paul Wood, 311–315. Malden, MA: Blackwell, [1920] 1992.

Raboy, Marc. *Marconi: The Man Who Networked the World*. Oxford: Oxford University Press, 2016.

Reichlin, Bruno. "The Axonometric as a Project: A Study of Alberto Sartoris." *Lotus International* 22 (1979): 83–93.

Reizman, Renée. "What Killed the Pay Phone." *The Atlantic* (February 2, 2017). Accessed June 25, 2022. https://www.theatlantic.com/technology/archive/2017/02/object-lesson-phone-booth/515385/.

Riley, Terence. *The Un-Private House*. New York: Museum of Modern Art, 1999.

Robida, Albert. *Le vingtième siècle: La vie électrique*. Paris: Librairie illustrée, 1893.

Rosenkranz, Karl. *Aesthetics of Ugliness: A Critical Edition*, edited and translated by Andrei Pop and Mechtild Widrich. New York: Bloomsbury, [1853] 2015.

Rouillard, Dominique, ed. *Imaginaires d'infrastructure*. Paris: L'Harmattan, 2009.

Rykwert, Joseph, Neil Leach and Robert Tavernor, eds. *Leon Battista Alberti, On the Art of Building in Ten Books*. Cambridge, MA: MIT Press, 1991.

Sandweiss, Martha A. *Picturing U.S. History*. Accessed November 13, 2019. https://picturinghistory.gc.cuny.edu/john-gast-american-progress-1872.

Sakarovitch, Joël. "Architecture et représentation: de la projection à la double projection." *In Extenso*, edited by Roger Laurent, no. 13 (1990): 63–101.

Schaeffer, Pierre. *Traité des objets musicaux: Essais interdisciplinaires*. Paris, Éditions du Seuil, 1966.

Schivelbusch, Wolfgang. *The Railway Journey: The Industrialization of Time and Space in the 19th Century*. Berkeley: University of California Press, 1986.

Scolari, Massimo. *Il disegno obliquo: una storia dell'antiprospettiva*. Venice: Marsilio Editori, 2005.

Segal, Howard P. *Technological Utopianism in American Culture*. Syracuse, NY: Syracuse University Press, 2005.

Segal, Howard P. "Mumford's Alternative to the Megamachine: Critical Utopianism, Regionalism, and Decentralization." In *Lewis Mumford: Public Intellectual*, edited by Thomas P. and Agatha C. Hughes, 100–109. New York / Oxford: Oxford University Press, 1990.

Siegert, Bernhard. "(Not) in Place: The Grid, or, Cultural Techniques of Ruling Spaces." In *Cultural Techniques: Grids, Filters, Doors, and Other Articulations of the Real*, 97–120. New York: Fordham University Press, 2015.

Simondon, Georges. *Du mode d'existence des objets techniques*. Paris: Aubier, [1958] 2015.

Sloterdijk, Peter. *Ecumes: Spheres III*. Translated by Olivier Mannoni. Paris: Hachette Littératures, 2005.

Stamp, Gavin. *Telephone Boxes*. London: Chatto & Windus, 1989.

Sterne, Jonathan. *The Audible Past*. Durham, NC: Duke University Press, 2003.
Sterne, Jonathan. "Sound Out of Time / Modernity's Echo." In *Turning the Century: Essays in Media and Cultural Studies*, edited by Carol A. Stabile, 9–30. Boulder, CO: Westview Press, 2000.
Sterne, Jonathan. "Transportation and Communication: Together as You've Always Wanted Them." In *Thinking with James Carey: Essays on Communications, Transportation, History*, edited by Jeremy Packer and Craig Robertson, 117–135. New York: Peter Lang, 2006.
Stilgoe, John R. *Outside Lies Magic*. New York: Walker & Company, 1998.
Sudhalter, Adrian. "Friedrichstrasse: The Contexts of an Image, 1922–1924." In *Mies van der Rohe: Montage Collage*, edited by Andreas Beitin, Wolf Eiermann and Brigitte Franzen, 68–85. London: Koenig, 2017.
Testavin, Eugène. "L'organisation actuelle du théâtrophone." *Annales des Postes, télégraphes et téléphones* 19, no. 1 (1930): 2–26.
Teyssot, Georges. "A Topology of Thresholds." *Home Culture* 2, no. 1 (2005): 89–116.
Teyssot, Georges. "Windows and Screens: A Topology of the Intimate and the Extimate." *Log*, no. 18 (2010): 75–88.
Thiermann Riesco, Alfredo. *Radio-Activities: Architecture and Broadcasting in Cold War Berlin*. Cambridge, MA: MIT Press, 2024.
Thiermann Riesco, Alfredo. "Radio Activities: The Architecture of Broadcasting in Cold War Berlin." PhD diss., Swiss Federal Institute of Technology Zurich, 2020.
Thompson, Emily. *The Soundscape of Modernity: Architectural Acoustics and the Culture of Listening in America, 1900–1933*. Cambridge, MA: MIT Press, 2002.
Thoreau, Henry David. *The Writings of Henry David Thoreau, Journal III, 1851–1852*. Boston: Houghton Mifflin Co., 1906.
Tsing, Anna. *Friction: An Ethnography of Global Connection*. Princeton: Princeton University Press, 2005.
Turgan, Julien. *Les grandes usines, études industrielles en France et à l'étranger*. Paris: Calmann Lévy, 1882.
Tyrwhitt, Jaqueline. "The Valley Section: Patrick Geddes's World Image." *Journal of the Town Planning Institute*, no. 37 (January 1951): 61–66.
Van der Marck, Jan. *Art by Telephone*. Chicago: Museum of Contemporary Art, 1969.
Van Drie, Melissa. "Devenir auditeur: une nouvelle conscience du son et de soi à travers l'expérience du théâtrophone (1881–1936)." In *Musique et enregistrement*, edited by Pierre-Henry Frangne and Hervé Lacombe, 55–76. Rennes: PUR, 2014.
Van Drie, Melissa. "Hearing through the Théâtrophone: Sonically Constructed Spaces and Embodied Listening in Late

Nineteenth-Century French Theatre." *Sound Effects: An Interdisciplinary Journal of Sound and Sound Experience* 5, no. 1 (2015): 73–90.

Van Drie, Melissa. "L'espace scénique du Théâtrophone (1881–1930) et la figure nouvelle du spectateur-auditeur." *Cahiers Création Arts et Patrimoines: Modèles et modalités de la transmission culturelle* 2 (2015): 41–68.

Varnelis, Kazys, and Anne Friedberg. "Place: The Networking of Public Space." In *Networked Publics*, edited by Kazys Varnelis. Cambridge, MA: MIT Press, 2008.

Venturi, Robert, Denise Scott Brown, and Steven Izenour. *Learning from Las Vegas: The Forgotten Symbolism of Architectural Form*. Cambridge, MA: MIT Press, [1972] 2001.

Virilio, Paul. *Ce qui arrive*. Paris: Éditions Galilée, 2002.

Virilio, Paul. *La vitesse de libération*. Paris: Galilée, 1995.

Von Moos, Stanislaus. *Esthétique industrielle*. Disentis: Pro Helvetia/ Editions Desertina, 1992.

Wajcman, Gérard. *Fenêtre: Chroniques du regard et de l'intime*. Lagrasse: Verdier, 2004.

Webster, Daniel. *Mr. Webster's address at the laying of the cornerstone of the addition to the Capitol, July 4th, 1851*. Washington: Gideon and Co. Printers, 1851.

Welter, Volker M. *Biopolis: Patrick Geddes and the City Life*. Cambridge, MA: MIT Press, 2002.

Wigley, Mark. *Buckminster Fuller Inc.: Architecture in the Age of Radio*. Zurich: Lars Müller, 2015.

Williams, Carlton D. "Acoustic Space." *Explorations: Studies in Culture and Communication*, no. 4 (February 1955): 15–20.

Williams, Raymond. *Television: Technology and Cultural Form*. London: Fontana, 1974.

Wojtowicz, Robert. *Lewis Mumford and American Modernism: Eutopian Theories for Architecture and Urban Planning*. Cambridge / New York: Cambridge University Press, 1996.

Wright, Frank Lloyd. *Broadacre City: A New Community Plan*. In *America Builds: Source Documents in American Architecture and Planning*, edited by Leland M. Roth. New York: Harper & Row, [1935] 1983.

Wright, Frank Lloyd. *The Disappearing City*. New York: W. F. Payson, 1932.

Wright, Frank Lloyd. *The Living City*. New York: Horizon Press, 1958.

Wright, Frank Lloyd. *When Democracy Builds*. Chicago: University of Chicago Press, 1945.

Index

Page numbers in italics refer to figures.

The MIT Press
Massachusetts Institute of Technology
77 Massachusetts Avenue, Cambridge, MA 02139
mitpress.mit.edu

The MIT Press would like to thank the anonymous peer reviewers who provided comments on drafts of this book. The generous work of academic experts is essential for establishing the authority and quality of our publications. We acknowledge with gratitude the contributions of these otherwise uncredited readers.

This book was set in Haultin by Jen Jackowitz. Printed and bound in the United States of America.

Library of Congress Cataloging-in-Publication Data

Names: Darò, Carlotta, author. | Hubert, Christian, 1948- translator.
Title: The architecture of the wire : infrastructures of telecommunication / Carlotta Darò ; translated by Christian Hubert.
Other titles: Paysage de lignes. English
Description: Cambridge, Massachusetts ; London, England : The MIT Press, [2025] | Translation of: Paysage de lignes. Genève : Métis Presses, 2022. | Includes bibliographical references and index.
Identifiers: LCCN 2024019129 | ISBN 9780262551632 (paperback)
Subjects: LCSH: Electric lines—Poles and towers—History. | Electric lines—Poles and towers—Pictorial works. | Telephone wire—History. | Architecture, Industrial.
Classification: LCC TK6351 .D3713 2025 | DDC 384.6—dc23/eng/20241126
LC record available at https://lccn.loc.gov/2024019129

10 9 8 7 6 5 4 3 2 1

EU product safety and compliance information contact is:
mitp-eu-gpsr@mit.edu

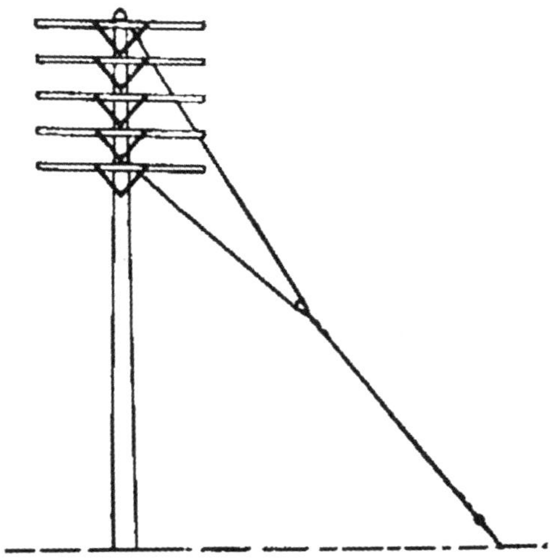